Data-driven Modelling of Wind Farm Flow Control Strategies

Nassir Cassamo • Jan-Willem van Wingerden

Data-driven Modelling of Wind Farm Flow Control Strategies

Springer

Nassir Cassamo
Vattenfall
Amsterdam, The Netherlands

Jan-Willem van Wingerden
Delft University of Technology
Delft, The Netherlands

ISBN 978-3-031-84115-6 ISBN 978-3-031-84116-3 (eBook)
https://doi.org/10.1007/978-3-031-84116-3

This Springer imprint is published by the registered company Springer Nature Switzerland AG
The registered company address is: Gewerbestrasse 11, 6330 Cham, Switzerland

"Your significance will remain forever obscure to you, but you may assume that you are fulfilling your role if you apply yourself to converting your experiences to the highest advantage of others" R. Buckminster Fuller

Preface

As the covid pandemic swept through Europe, my goodreads book wish list grew longer. As the lockdowns started becoming a harsh reality, many of these books found their way to my home. One of these books was entitled *Range: Why Generalists Triumph in a Specialized World.* In this book, David Epstein, the author, made the case that generalists, this breed that purposefully chooses to bridge gaps across domains of knowledge, as opposed to experts who deep dive into one specific subject, would more easily thrive in the modern world. The following quote in the book's back cover captures this philosophy well: *The most impactful inventors cross domains rather than deepen their knowledge in a single area. As experts silo themselves further while computers master more of the skills once reserved for highly focused humans, people who think broadly and embrace diverse experiences and perspectives will increasingly thrive.* My thinking was, at the time, incredibly aligned to the author's core ideas. It had always fascinated me when different fields of science came together, instilling in me a felling of limitlessness. Creatively merging two subjects, in my mind, could open endless possibilities. Boredom was not part of this picture.

This book represents one step in my life journey as a generalist. Its title, *Data Driven Modelling of Wind Farm Control Strategies*, brings many fields of knowledge together such as wind farm control and data driven modelling. One could argue that wind farm control is a subject that arises from the crossing of two other areas, control systems and fluid mechanics. Data-driven modelling is similarly built on top of other fields such as algebra and non-linear dynamical systems theory. As the proud generalist I consider myself to be this book represents an in depth-contribution of the result of crossing these two subjects.

I will feel that this book has accomplished its purpose if you, the reader, by means of it, have achieved three things at the end. First, gain a fundamental understanding of the two core subjects in this book: wind farm control and data driven modelling. The latter two are vast fields of expertise and the contents that will follow represent a comprehensive introduction. Relevant topics and recent developments of these two subjects are discussed. The selected information included throughout the book provides a gateway for the reader so that she or he can get familiarized with these fields and deepen her or his knowledge with further dedicated material. Second, gain practical experience in the application of data-driven algorithms within the context of wind farm control. Code snippets with examples are given throughout the book and the original datasets used can be also accessed so that the algorithms can be tested. The learning experience is intended to be practical and hands-on. Third, innovate. Different variations of the algorithms can be tested or, for example, an implementation of the existing algorithms can be performed in a different programming language. All of these are encouraged and the learning-by-doing philosophy is at the heart of this book.

This book was written to support the generalist's path, which is often a hard one to walk. It is a cocktail of steep learning curves, failures and is sprinkled with a pinch of imposter syndrome. That is the one I chose to walk at the beginning of my thesis and would not have been successful if not for the guidance of my kind supervisor and co-author Jan-Willem van Wingerden. More than a mere literary instrument to pass along knowledge, this book aims to equip the reader with tools to develop new ideas applicable to data-driven modelling of wind farm control strategies and other fields as well. Although it has become clear that my inclination falls towards the generalist as a means to drive innovation, one must never disregard or undermine the role of the expert. The relationship between the two is of a beautiful symbiosis. As a generalist, I gain immense satisfaction from absorbing colossal amounts of new information on topics I am passionate about and generating new ideas from merging distinct topics. This is only possible due to the ingenious breakthroughs that dedicated individuals have achieved throughout decades, if not centuries, of in-depth research. As these individuals explore and create new mathematical algorithms by leveraging their in-depth expertise, the generalists can then apply them to other fields and further explore their capabilities. In case you are new to these two subjects—wind farm control and data-driven modelling—do not let the novelty or current absence of knowledge impede you to start the learning journey. Hopefully the contents of this book will help you navigate the path ahead.

The contents of this book are solely the product of my personal research, thoughts and experiences. They do not in any way reflect the views, opinions or policies of my employer, Vattenfall. Furthermore, the ideas and concepts presented within these pages are not related to the work undertaken during my tenure at Vattenfall. Rather, they represent the culmination of endeavours predating my current employment. I am solely responsible for the content herein and any discrepancies or misinterpretations should not be attributed to my employer.

Amsterdam, The Netherlands — Nassir Cassamo
Delft, The Netherlands — Jan-Willem van Wingerden

The original version of the book has been revised. A correction to this book can be found at https://doi.org/10.1007/978-3-031-84116-3.

Contents

Acronyms

ABL	Atmospheric Boundary Layer
ADM	Actuator Disk Model
AI	Artificial Intelligence
AIC	Axial Induction Control
ALM	Actuator Line Model
CFD	Computational Fluid Dynamics
DIC	Dynamic Induction Control
DIO	Direct Input Output
DIPC	Dynamic Individual Pitch Control
DMD	Dynamic Mode Decomposition
DMDc	Dynamic Mode Decomposition with Control
DOF	Degrees Of Freedom
EDMD	Extended Dynamic Mode Decomposition
EIODMD	Extended Input Output Dynamic Mode Decomposition
FAST	Fatigue Aerodynamics Structures and Turbulence
HAWT	Horizontal Axis Wind Turbine
IIO	Increment Input Output
IODMD	Input-Output Dynamic Mode Decomposition
IPC	Individual Pitch Control
IPCC	Intergovernmental Panel on Climate Change
LCOE	Levelized Cost of Electricity
LES	Large Eddy Simulation
LPV	Linear Parameter Varying
LQG	Linear Quadratic Gaussian
LQR	Linear Quadratic Regulator
LTI	Linear Time Invariant
MBPC	Model Based Predictive Control
MPC	Model Predictive Control
NPDE	Nonlinear Partial Differential Equations
NREL	National Renewable Energy Laboratory
NRMSE	Normalized Root Mean Squared Error

NS	Navier Stokes
POD	Proper Orthogonal Decomposition
QP	Quadratic Programming
RANS	Reynold Averaged Navier Stokes
RBS	Randomized Binary Signals
ROM	Reduced Order Model
SOWFA	Simulator for Offshore Wind Farm Applications
SVD	Singular Value Decomposition
TEM	Topical Expert Meeting
VAF	Variance Accounted For
WECS	Wind Energy Conversion System
WRC	Wake Redirection Control

Chapter 1
Introduction

This first introductory chapter provides the motivations to study the topics that are discussed throughout this book. The incentives and benefits of applying data-driven techniques to systems of high complexity, such as wind farms, are discussed. Other motivating forces are introduced. The text continues by describing the objectives of this book as a means to provide the theoretical and mostly practical tools so that the reader can apply these to similar research. The structure of the following chapters is then presented.

1.1 Motivation

Wind energy offers a potential to reduce carbon emissions and mitigate worldwide climate change. As this technology is harvested throughout the world, many challenges arise when attempting to extract the most benefit out of wind turbines.

In order to be successful in this task, it becomes essential to derive accurate models which can describe the power production of wind turbines, so that engineers may predict and optimize renewable power production. As wind turbines are placed together offshore in wind farms, models should then be able to account for the complex phenomena that take place, such as the aerodynamic interaction between turbines. This becomes challenging, as such systems are usually high dimensional and governed by non-linear equations which possess no known analytical solution as of today.

As a motivating example, consider wind farms Borselle 1 and 2. This 752MW wind farm, recently installed in 2020, in the Dutch North Sea, is comprised of 94 wind turbines distributed over a site area of 112 km^2 [1]. Deriving a model which can reproduce the behaviour of this wind farm is naturally challenging. Ideally engineers strive to find simple models, as they are easier to manipulate for the purposes of optimization.

N. Cassamo, J.-W. van Wingerden, *Data-driven Modelling of Wind Farm Flow Control Strategies*, https://doi.org/10.1007/978-3-031-84116-3_1

The high dimensions of wind farms such as the Borselle 1 and 2 require complex models to accurately reproduce the inherent dynamics (aerodynamic interaction between turbines, power production of individual turbines, loads experienced by different turbines, etc.). The complexity of such models may be a bottleneck to perform optimization, especially if it is intended to be in real time. There exists an incentive to study and explore various modelling techniques so that simple models can be used to optimize wind farm operating characteristics.

Various motivating forces to study power optimization problems in wind farms exist, which emerge as a means to provide answers to the current challenges.

A compelling argument to further investigate new data-driven approaches to model the behaviour of wind farms is the increasing interest in understanding the complex dynamics in wind farms that have an impact in power output of wind assets.

In late 2019, a company announcement by Ørsted [2], a Danish multinational power company based in Denmark, presented an update on its long-term financial targets, focusing on three factors which were claimed of having added pressure on Ørsted's long-term targets. The first factor was related to offshore production forecasts, where an internal project led to the conclusion that (...) *our current portfolio production forecasts underestimate the negative impact of two effects across our asset portfolio, i.e. the blockage effect and the wake effect.*. The key message with this small excerpt is that these phenomena present in wind farms are in need to be investigated and better understood. The final goal is to design more accurate models that can take these effects into account so that production forecasts can be more precise.

The blockage effect and the wake effect were two of the effects pointed out as not being taken into consideration sufficiently. The wake effect is at the core of this book. As the wind is extracted by turbines positioned in the first row of turbines of a wind farm, the wind will slow down, making less energy available for downstream turbines. Not only were the wakes within a wind farm pointed out as an effect, but also the wakes between neighbouring wind farms.

The blockage effect, in turn, refers to the redistribution of energy in the upstream wind throughout the farm and its surroundings. This is caused by the thrust force exerted by turbines when extracting energy from the wind. The broadly accepted industrial consensus is that there is a reduction of the upstream wind speed; however it is not yet clear how to consider blockage effects' impacts downstream [3].

The flow will decelerate and deviate as it approaches the wind turbine. For an isolated wind turbine, this effect is better understood and formulas which quantify the decrease in wind speed exist [4]. However, for large wind farms, this effect becomes increasingly more complex to model [5]. It is usually referred to as Global Blockage, given that it is a combined effect of all turbines operating in a wind farm.

Investigating and tackling wake interactions within wind farms, between farms and global blockage effects is of extreme relevance at the present moment. As a motivating example, take Wake Adapt, a product developed by Siemens Gamesa Renewable Energy (SGRE), a company in the renewables energy sector, attempting to mitigate wake effects. According to SGRE, the new wind farm control product can determine the direction of wind reaching downstream turbines and up to a gain of 1% in Annual Energy Production [6]. Global Blockage Effects are also being investigated. A consortium formed by companies in the offshore wind energy industry and led by RWE, a German company in the energy sector, is conducting one of its kind measurement campaign to quantify wind farm blockage under real offshore conditions [7]. The project aims to not only measure the accelerations and decelerations of the wind and their sensitivity to atmospheric parameters, but also to build a consensus across industry around the modelling principles of blockage [3].

> Better understanding complex phenomena in wind farms is a core issue in the everyday life of many companies in the energy sector. Such knowledge drives consensus and promotes enhanced modelling approaches across industry. Data-driven techniques offer an alternative approach and have the possibility of adding relevant insights to the discussion.

Other motivating forces to study data-driven approaches for the purposes of modelling and strategies to mitigate wake effects are:

1. *Offshore wind energy production is rapidly growing and technologies are developing:* Renewable energy production has increased substantially in the last years, with a 5–6 fold increase since the 1960s [8]. Wind energy is one of the fast growing technologies, representing 19% of the total renewable energy production in 2018, compared to 8% in 2010.

   The total installed wind power capacity in Europe has been steadily increasing since 2009, as shown by Wind Europe, a European agency [9]. Statistics indicate an increase from 77 GW (75 onshore and 2 offshore) of cumulative wind power capacity in 2009 to 205 GW (183 onshore and 22 offshore) in 2019. In 2019, Europe installed 15.4 GW of new wind power capacity, 28% more than 2018. Onshore wind was 76% of the new installations with 11.7 GW, and offshore installations were a record 3.6 GW. In 2020, Europe had a total installed power capacity of 220 GW (194 GW onshore and 25 GW offshore), according to a report from Wind Europe [10]. The trend for going offshore is explained by its advantages, namely the high availability of resources—wind speeds are higher—improved social acceptability, lower environmental impacts such as noise and potential for reduced Levelized Cost of electricity (LCOE).[1]

The trend for growth also applies to wind turbines and wind farms. Installed offshore wind turbines registered an average rated capacity[2] of 8.2 MW in 2020. Turbine orders in 2020, however, already show a trend towards the next generation in size, ranging from 10 to 13 MW for projects coming online after 2022, as indicated in [11]. The same applies to wind farms, with the average wind farm in 2020 of 788 MW (26% more than in 2019).

2. *It is an active and promising research topic which tackles wind farms intrinsic challenges:* In spite of the economic advantages of sitting wind turbines together in wind farms, the aerodynamic interaction between turbines may have a negative effect on the total electrical power production and may increase the loads experienced by turbines. The aerodynamic interaction effects are caused by the turbine wakes, which are the flow structures that form behind each turbine. Mitigation of these effects can be achieved either by optimizing the wind farm layout and/or by using wind plant control techniques during the operation of the wind plant. The latter concept aims at improving the performance of the wind plant as a whole by maximizing the power output of the plant instead of each individual turbine, which can most often be suboptimal.

   It was firstly introduced in 1988 in [12], showing that slight increases in the overall energy capture are possible. Since then, various control strategies have been studied in depth, resorting to simulations, wind tunnel experiments and real wind farm tests. The topic of wind farm flow control has also been identified as a core component of the grand challenges in wind energy research needs for a successful global energy transition, mentioned to offer opportunities for enhanced production, lower maintenance costs and increased controllability required by the overall energy system [13]. The authors in [14] further identify four main areas where open scientific challenges and opportunities still exist in the field of wind farm flow control. The first relates with the physics of the atmosphere and physics induced by the flow control strategies, stating that the effects of certain conditions on the shape of the wake are not yet entirely understood. The others refer to algorithms and Artificial Intelligence, validation and implementation and co-design, i.e. incorporating farm control in the plant layout optimization process.

   More recently, dynamic wind farm control has gained interest in the scientific field. This is based on varying the control input through time, which differs from steady-state optimal control. The underlying hypothesis is that time-varying

[1] Represents the average revenue per unit of electricity generated that would be required to recover the costs of building and operating a generating plant (during an assumed financial life and duty cycle).

[2] Rated capacity is the maximum amount of power a wind turbine is designed to generate under specific conditions such as high wind speeds.

inputs are able to purposely influence the inherently dynamic nature of the wind. The potential of such strategies has been shown in high fidelity simulations and more recently, in 2019, wind tunnel experiments [15, 16]. Moreover, new control strategies are being developed based on these new novel concepts, such as the helix approach, where the wake presents a helicoidal shape [17]. Understanding optimal dynamical controlling techniques is still an on-going research topic and a fundamental one in order to achieve meaningful power production levels. Data-driven modelling techniques provide a means to better understand the underlying physical and atmospheric phenomena that lead to specific modification in the flow and subsequent power increases which originate from dynamic actuation.

3. *Data-driven techniques provide a possible route to the modelling of complex systems:* Systems which are of interest to study can be complex, in the sense that they are high dimensional and exhibit non-linear behaviour. A wind farm falls into such category, and although the governing equations are known, the underlying mathematical complexity makes them undesirable to work with. In other systems the underlying governing equations may even be unknown, making the task of modelling more complicated. An example of such system is the brain, where data-driven techniques play a relevant role in understanding key patterns in large-scale recordings of neural activity.

   Data-driven techniques offer an alternative route for the modelling of such complex systems which is equation free. This means that as opposed to starting with the knowledge of the equations governing a certain system, data is acquired and used to derive a model of the system. These techniques are specially relevant in the present time, where there is an unprecedented amount of data available as well as computational resources to make the most out of it.

4. *Wind energy production is a pivotal step in preventing global warming complications:* The Summary for Policymakers presents the key findings of the Special Report produced in 2018 by the Intergovernmental Panel on Climate Change (IPCC). The latter begins by noting that human activities are estimated to have caused approximately 1.0 °C of global warming above pre-industrial levels and that it is *likely* to reach 1.5 °C between 2030 and 2052, if it continues to increase at the current rate. The most recent summary from 2022 further states that if the 1.5 °C threshold is reached, multiple climate hazards, such as extreme weather events, will unavoidably multiply. The document further states that near term actions to limit global warming to levels below 1.5 °C would *substantially reduce projected losses and damages related to climate change in human systems and ecosystems* (...), despite not being able to eliminate them all.

   The knowledge that global warming presents a serious issue has been around for more than two decades. The Kyoto Protocol, launched in 1995, was the first international tool focused on greenhouse gas mitigation with initially 37 countries pledging to reduce their emissions by 5% compared to the 1990 baseline in the period 2008–2012 [18]. The Paris Agreement is a more recent effort, bringing all nations together with a clear aim: to strengthen the global response to the threat of climate change by keeping the global temperature rise this century well below 2 °C above pre-industrial levels and to pursue the efforts to limit temperature increase even further to 1.5 °C. These values are based on the Special Report findings, which highlights that a number of climate changes could be avoided by limiting global warming to 1.5 °C compared to 2 °C, or more.

In order to meet the agreements' targets and circumvent disastrous scenarios, countries started to adopt new measures, which place great emphasis on energy policies. The Kyoto Protocol, for example, states in Article 2 that countries shall implement and/or further elaborate policies such as *research on, and promotion, development and increased use of new and renewable forms of energy, of carbon dioxide sequestration technologies and of advanced and innovative environmentally sound technologies* [19]. More recently, the European Union (EU) has pledged to become carbon neutral by 2050, i.e. an economy with net-zero[3] greenhouse gas (GHG) emissions. This commitment is at the core of the European Green Deal, a set of initiatives and policies taken within the group of 27 countries which aim at curbing carbon emissions and achieving economic growth whilst maintaining resource use at minimum levels.

Learning about wind farm control techniques and their potential to increase Annual Energy Production (AEP) of wind farm offshore assets is a step towards stimulating the adoption of these technologies and contributing to a carbon neutral future.

On top of industrial interest in having a better grasp of the phenomena that drive power production in wind farms, other motivating forces to focus on the contents of this book exist. These include, for example, growing interest in techniques to mitigate wake effects—a continuously growing and maturing field—the blooming of the offshore wind energy industry, the vast amount of data available and experimented via new algorithms and the fact that wind energy and wind farm control are part of the solution to achieve carbon neutrality in the EU.

1.2 Objectives

This book has three objectives:

1. *Introduce the topic of wind farm control:* The topic of wind farm control is an active and growing field of research. This book introduces the topic, covering the

[3] Net zero in the sense that GHG emissions are offset by removing the same equivalent amount from the atmosphere.

main control strategies to mitigate wake effects and state of the art developments. The literature is vast and continuously growing. The reader is encouraged to complement the contents of this book with additional material, which is usually made available in the form or conference papers and journal articles.

2. *Illustrate the application of the latest data-driven algorithms in the context of wind farm control:* Data-driven techniques have a wide range of applications and are suitable to study different systems. In this book the latest developments in data-driven modelling methodologies are discussed and applied to different wind farm control strategies. Different variations of the techniques are showcased, so the reader has an overview of the range of possibilities in using such techniques.

3. *Empower readers with the necessary tools:* All necessary material is fully made available to allow the reader to maximize the learning about the subject and to inspire the reader to apply these techniques to their own research projects. Data from high fidelity simulations used throughout the book as well as the developed MATLAB code can be found at https://github.com/nassircassamo/IODMD_SOWFA. The most important algorithms are shown and discussed. As stated earlier, learning by doing is at the core of this book. The datasets are shared in the hope that the reader will explore them, apply variations of the algorithms and test new variations. The datasets can be downloaded in [20] and [21], for pitch- and yaw-based control, respectively.

1.3 Outline

This book is divided into seven distinct chapters, with the first one being this introductory chapter. The remaining ones are the following:

2. *Wind Turbine and Wind Farm Control*: The theoretical backgrounds regarding wind turbine and wind farm control are discussed in the first chapter. These include the basic operating settings available in wind turbines for control, as well as typical terms used in the scientific field. In addition, the main wind farm control strategies are introduced, with accompanying explanations of the physics involved.

3. *Data-Driven Modelling of Wind Farm Control strategies*: The second chapter starts by discussing the task of modelling a wind farm and the inherently mathematical complexity involved in the classical physics-based approach. The data-driven approach is offered as an alternative. In other words, as opposed to resorting to the underlying mathematical equations to derive a model, only data—in the approach explored in depth in this book, measurements taken from a simulation—are used to derive a model. This chapter introduces the core

technique explored within this book, a variant of the famous Dynamic Mode Decomposition (DMD) algorithm. Fundamental mathematical concepts from dynamical systems theory which motivate the variations of the DMD algorithms explored in this work are also discussed in this section.

4. *Methodology for Modelling Wind Farm Control Strategies*: In Chaps. 5 and 6, the techniques introduced in Chap. 3 are applied to two distinct wind farm control strategies. In other words, Chap. 3 describes the mathematical details of these techniques, whereas Chaps. 5 and 6 outline a practical application to two distinct wake control technologies. This chapter focuses on describing the procedure of model identification and validation which will later be used.

5. *Data-Driven Modelling of Wake Redirection Control*: The fifth chapter deep dives into the application of DMD related algorithms to compute a model for Wake Redirection Control, the first wind farm control strategy which will be explored. Data from the simulation is first evaluated to provide the reader with a concrete and tangible illustration of the workings of this control strategy. Different models are compared and the best performing model is explored in greater detail.

6. *Data-Driven Modelling of Axial Induction Control*: The sixth chapter is similar to the fifth one; however this one focuses on Axial Induction Control, a different wind farm control strategy. Furthermore, it explores other types of DMD related algorithms which are not explored in Chap. 5.

7. *Advanced Topics*: The seventh and last chapter explores two pertinent topics related with the simulation and post-processing of wind farm data-driven models. The first is the estimation of initial conditions to perform linear simulations, where two alternative methods are investigated. The second is related to the regularization approach to force certain models to become stable.

As this book combines concepts from different domains of science—linear algebra, systems identification, dynamical systems theory, fluid mechanics, wind turbine and wind farm control—the most relevant mathematical tools are described in Annex A.2 and short footnotes are occasional used to clarify simpler concepts. The following notation is used throughout this book: Vectors and matrices are printed in a boldface font (e.g. $\mathbf{A}$, $\boldsymbol{\Sigma}$). Vectors are usually denoted by lower cases (e.g. $\boldsymbol{g}$) and matrices by a capital letter (e.g. $\mathbf{G}$).

Whether the reader is a wind farm control enthusiast attempting to gain insights into wind farm control strategies, or a curious aficionado of the latest data-driven techniques, the open disclosure of this material intends to allow for reproducible research and promote the learning and implementation of these mathematical techniques.

References

1. Tom Christiansen Ørsted Media Relations. *Ørsted's first Dutch offshore wind farm fully commissioned.* http://www.wpcentral.com/ie9-windows-phone-7-adobe-flash-demos-and-development-videos
2. Martin Barlebo Ørsted Media Relations. *Ørsted presents update on its long-term financial targets.* https://orsted.com/en/company-announcement-list/2019/10/1937002
3. Carbon Trust. *Global Blockage Effect in Offshore Wind (GloBE).* https://www.carbontrust.com/our-projects/large-scale-rd-projects-offshore-wind/global-blockage-effect-in-offshore-wind-globe
4. A. Segalini, J.-Å. Dahlberg, Global blockage effects in wind farms. J. Phys. Conf. Ser. **1256**(1), 012021 (2019). https://doi.org/10.1088/1742-6596/1256/1/012021
5. M. Popescu, T. Flåtten, A study of blockage effects at the wind turbine and wind farm scales. Energies **14**(19) (2021). ISSN: 1996-1073. https://doi.org/10.3390/en14196124. https://www.mdpi.com/1996-1073/14/19/6124
6. Veronica Diaz Siemens Gamesa Renewable Energy. *Press release: Siemens Gamesa now able to actively dictate wind flow at offshore wind locations.* https://www.siemensgamesa.com/en-int/-/media/siemensgamesa/downloads/en/newsroom/2019/11/press-release-siemens-gamesa-wake-adapt-en.pdf
7. Carbon Trust, *Offshore wind industry consortium gears up to conduct first of a kind full-scale measurement campaign focused on Global Blockage Effect.* https://www.carbontrust.com/news-and-events/news/offshore-wind-global-blockage-effect
8. H. Ritchie, Renewable energy, in *Our World in Data* (2017)
9. I. Komusanac, G. Brindley, D. Fraile, *Wind Energy in Europe: Trends and Statistics.* Technical Report Global Wind Energy Council (2020)
10. I. Komusanac et al., *Wind energy in Europe: 2020 Statistics and the outlook for 2021-2025.* Technical Report. WindEurope (2021)
11. L. Ramírez, D. Fraile, G. Brindley, *Offshore Wind in Europe: key trends and statistics 2020.* Technical Report. WindEurope (2021)
12. M. Steinbuch et al., Optimal control of wind power plants. English. J. Wind Eng. Ind. Aerodyn. **27**(1–3), 237–246 (1988)
13. P. Veers et al., Grand challenges: wind energy research needs for a global energy transition. Wind Energy Sci. **7**(6), 2491–2496 (2022). https://doi.org/10.5194/wes-7-2491-2022. https://wes.copernicus.org/articles/7/2491/2022/
14. J. Meyers et al., Wind farm flow control: prospects and challenges. Wind Energy Sci. **7**(6), 2271–2306 (2022). https://doi.org/10.5194/wes-7-2271-2022. https://wes.copernicus.org/articles/7/2271/2022/
15. J.A. Frederik et al., Periodic dynamic induction control of wind farms: proving the potential in simulations and wind tunnel experiments. Wind Energy Sci. Discuss. **5**, 1–18 (2019)
16. D. van der Hoek et al., Experimental analysis of the effect of dynamic induction control on a wind turbine wake. Wind Energy Sci. **7**(3), 1305–1320 (2022). https://doi.org/10.5194/wes-7-1305-2022. https://wes.copernicus.org/articles/7/1305/2022/
17. J.A. Frederik et al., The helix approach: using dynamic individual pitch control to enhance wake mixing in wind farms. Wind Energy **23**(8), 1739–1751 (2020)
18. R. Morel, I. Shishlov, *Ex-post evaluation of the Kyoto Protocol: four key lessons for the 2015 Paris Agreement.* Technical Report A Caisse des Dépôts et Consignations (2014)
19. United Nations, *Kyoto Protocol to the United Nations framework convention on climate change* (1998)
20. N. Cassamo, J.-W. van Wingerden, *Dataset used in article: On the Potential of Reduced Order Models for Wind Farm Control: A Koop-man Dynamic Mode Decomposition Approach.* Zenodo (2020). https://doi.org/10.5281/zenodo.8287316
21. N. Cassamo, J.-W. van Wingerden, *Dataset used in article: Model Predictive Control for Wake Redirection in Wind Farms: a Koop-man Dynamic Mode Decomposition Approach.* Zenodo (2021). https://doi.org/10.5281/zenodo.8287375

Chapter 2
Wind Turbine and Wind Farm Control

This chapter introduces the topics of wind turbine and wind farm control. The main objectives of controlling a wind turbine and farm are introduced, as well as the available control variables for this purpose. Different wind farm control strategies can be devised by making use of the control variables at the turbine level. The former are introduced and discussed in this chapter.

2.1 Wind Turbines

2.1.1 Wind Turbine Components and Mechanisms

A wind turbine consists of a mechanical device specifically designed to extract kinetic energy from the wind and convert part of it into useful mechanical energy. The most commonly produced and used wind turbine is the Horizontal Axis Wind Turbine (HAWT). The main component of a HAWT is its rotor equipped with three rotor blades, which is attached to the generator through a drivetrain. The generator and drivetrain are housed in the nacelle, which is supported by a tower, as schematically represented in Fig. 2.1.

2.1.2 Wind Turbine Power Extraction

The rotor blades convert the momentum of a wind field passing the rotor plane into aerodynamic forces that drive the rotor. The drivetrain transfers the aerodynamic torque from the rotor to the generator shaft, directly or through a gearbox. The initial kinetic energy in the wind is then converted into electrical power.

N. Cassamo, J.-W. van Wingerden, *Data-driven Modelling of Wind Farm Flow Control Strategies*, https://doi.org/10.1007/978-3-031-84116-3_2

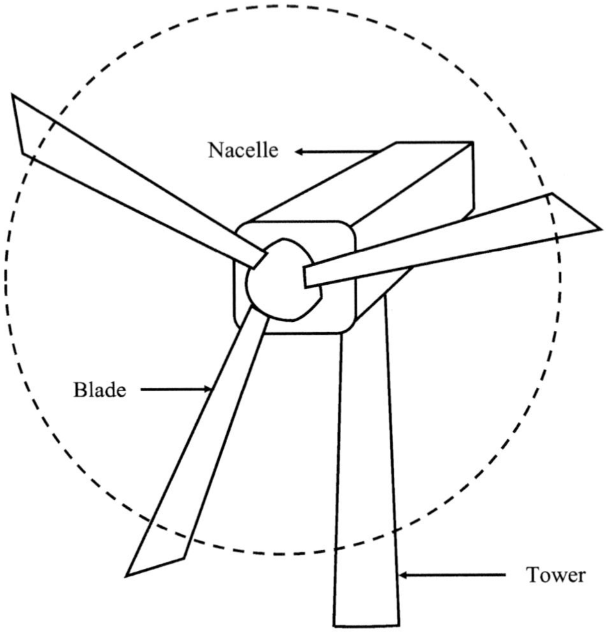

Fig. 2.1 Horizontal axis wind turbine (HAWT) main components. These include the three wind turbine blades, tower and nacelle

The available power in the wind due to its kinetic energy can be derived by evaluating the mass of wind flowing through an imaginary rotor disk of area A, which is swept by the rotor blades, and it is given by a well-known expression [1]:

$$P_w = \frac{1}{2}\rho A U_\infty^3 \tag{2.1}$$

where P_w [W] is the instantaneous kinetic power of the wind available at surface of area A [m^2], ρ [$\mathrm{kg} \cdot \mathrm{m}^{-3}$] is the air density which is assumed to be constant and U_∞ [$\mathrm{m} \cdot \mathrm{s}^{-1}$] the free stream wind velocity.

To evaluate the rotor performance in extracting energy from wind, the power coefficient C_P is used. It is defined as the dimensionless ratio of the power extracted by the rotor to the kinetic available power in the wind:

$$C_P = \frac{P}{\frac{1}{2}\rho A U_\infty^3} \tag{2.2}$$

As the air passes through the rotor disc, it proceeds downstream with reduced speed, a natural consequence of the extraction of energy from the wind. This region of the flow downstream of the turbine is called the wake, which will be explored in further depth in the following chapters.

2.1.3 The Axial Induction Factor

A measure of interest to characterize the drop in wind velocity between the free stream and rotor plane is the *axial induction factor* a, which is defined as the ratio of the difference between the free wind velocity U_∞ and the wind velocity at the rotor U_r to U_∞, represented in Eq. 2.3:

$$a = \frac{U_\infty - U_r}{U_\infty} \tag{2.3}$$

allowing to relate the wind speed at the rotor U_r with the free wind velocity U_∞:

$$U_r = U_\infty(1 - a) \tag{2.4}$$

The power coefficient can be written in terms of the axial induction factor by means of a simple model based on linear momentum theory generally attributed to Albert Betz in 1926. The model assumptions and derivation can be analysed in detail in [2]. The power coefficient is then rewritten as

$$C_P = 4a(1 - a)^2 \tag{2.5}$$

Taking the derivative of Eq. (2.5) with respect to a and equalling to zero, the maximum value of the power coefficient is found. In light of the underlying assumptions of the model, the maximum value of C_P occurs for $a = \frac{1}{3}$ and is equal to $C_{P_{\max}} = \frac{16}{27} = 0.5926$.

The $C_{P_{\max}}$ limit is famously known as the Betz limit. It states that, given the laws of physics, there exists a physical limit for the power a wind turbine can capture. Furthermore, the relation in (2.4) and the result for a of the Betz limit indicate that if an ideal rotor were designed and operated such that the wind speed at the rotor were 2/3 of the free stream wind speed, then it would be operating at the point of maximum power production.

The task is then to modify the wind turbine's settings such that the operating point of maximum power production is reached.

2.1.4 Wind Turbine Loading

The model based on linear momentum theory used for the analysis of power production can also be used to evaluate the force T of the wind on the wind turbine.

This is referred to as the thrust force, and it can be adimensionalized to give a coefficient of thrust C_T:

$$C_T = \frac{T}{\frac{1}{2}\rho A U_\infty^2} \tag{2.6}$$

and expressed by the axial induction factor (for full derivation, please see [1]): $C_T = 4a(1-a)$. The thrust force is a variable of interest, as one is not only interested in maximizing power production, but also in minimizing loading to increase the wind turbine's lifetime.

2.2 Wind Turbine Control

A Wind Energy Conversion System (WECS) connected to electric power networks—where a HAWT is included—must be designed to minimize the cost of supplied energy, whilst simultaneously ensuring safety, noise and power quality standards. This task, in turn, involves balancing a series of objectives, which must be pursued collectively. They include [3]:

1. *Energy capture*: Maximization of energy capture for certain inflow conditions (wind speed, wind direction, turbulence intensity) taking into account constraints, for example on maximum rotor speed due to safety issues or maximum noise constraints.

2. *Mechanical loads*: Preventing WECSs from excessive dynamical loads, both static (related to wind mean wind speed) and dynamic (related to wind speed spatio-temporal distribution over the rotor area).

3. *Power quality*: Power quality refers to the reliability and stability of the power supplied by the wind turbine. This relates to how well the power supply is able to maintain a consistent voltage, frequency and other desired characteristics. The task of controlling a wind turbine encompasses ensuring the generated power complies with pre-defined quality standards.

To guarantee that the above objectives are met, it is essential to control power production and loads exerted on the wind turbine. For such purpose, a number of Degrees of Freedom (DOF), also referred to as control variables, are typically available. These are depicted in Fig. 2.2.

The control variables available in typical commercial wind turbines are:

1. *Blade pitch angle* $\boldsymbol{\theta}$: The rotor blades can rotate, with their axis of rotation aligned with the blades, using hydraulic actuators or servo pitch motors. Changing the pitch angle directly changes the angle of attack along the different sections of the blade. In turn, such change affects the lift and drag forces so that in the end, the rotor speed and generated power can be controlled.

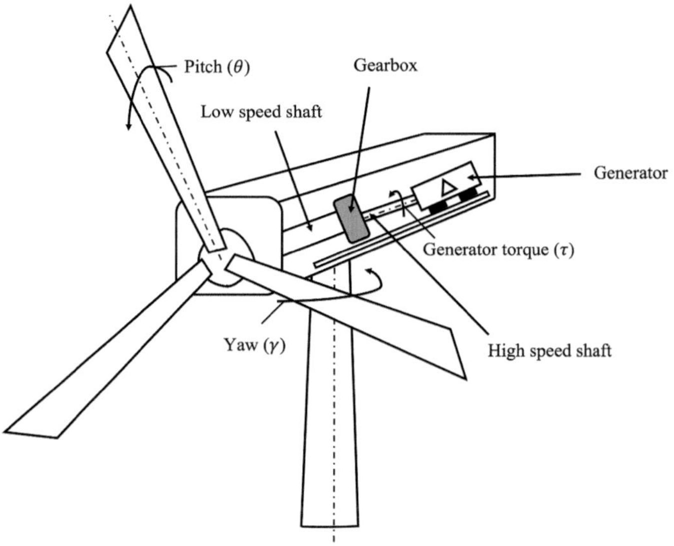

Fig. 2.2 Horizontal Axis Wind Turbine (HAWT) degrees of freedom available for the purposes of wind turbine control: blade pitch angle, nacelle yaw angle and generator torque

2. *Generator torque* $\boldsymbol{\tau}$: Torque control is used to control the power capture. Generator torque is regulated for that purpose either by design characteristics of the grid-connected generator or independently controlled with the use of power electronic converters.

3. *Yaw angle* $\boldsymbol{\gamma}$: The nacelle can rotate, with its axis of rotation aligned with the tower, by using a yaw motor. The yaw angle is defined as the angle between the axial rotor axis and the incoming wind direction. Yaw control allows to modify the angle between the rotor plane and the incoming wind direction.

The most common types of wind turbines are HAWT, although Vertical Axis Wind Turbines (VAWT) also exist. In addition, the most typically used are variable speed wind turbines that make use of control variables such as pitch and torque to regulate the turbines operation and extract the most energy out of the wind. In such turbines, three operating regions are usually distinguishable.

The different control variables play a fundamental role, and their usage defines distinct operating regions for the wind turbine functioning. For variable speed wind turbines, three regions are distinguished—represented in Fig. 2.3—each with its

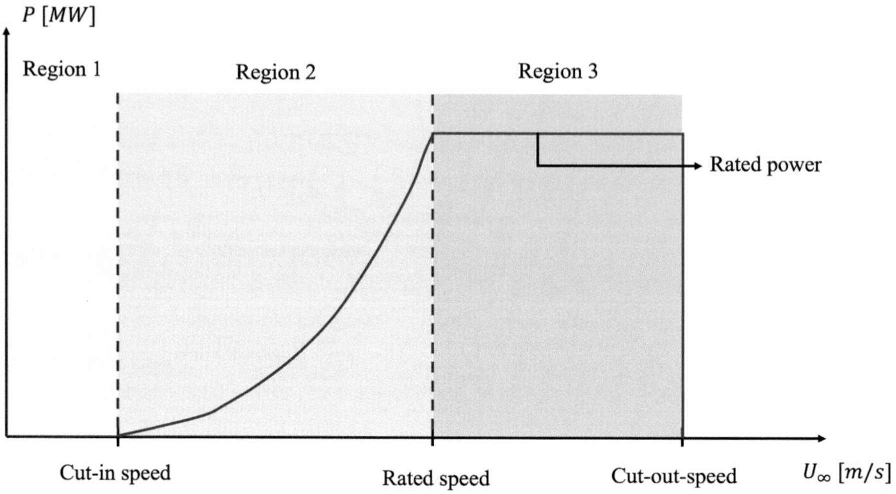

Fig. 2.3 Wind turbine power curve, different operating regions and main regulation strategies

control strategy [the rating (maximum power production point) of the turbine arises from a compromise between available energy and manufacturing costs]:

- *Region 1*: Below cut-in speed, the available wind energy is too low to compensate for the operation costs and losses. The cut-in speed is therefore defined as the minimum wind speed at which a wind turbine begins to generate electricity.

- *Region 2*: Below rated speed U_r torque control is employed, and the blades pitch angle remains constant. In this region, the so-called (optimal) λ^* Tip-Speed-Ratio (TSR) is tracked by varying the generator torque. The TSR is defined as $\lambda = \frac{\Omega R}{U}$, where Ω is the rotor speed, R is the rotor radius and U the wind speed. Below rated wind speed, i.e. $U < U_r$, this ensures that for the various wind speeds the turbine is operating with maximum aerodynamic efficiency.

- *Region 3*: Above rated speed, pitch control is employed. The rotor's blades pitch angle is varied so that an optimal rotor speed Ω^* is maintained and constant power is achieved. Above cut-out wind speed, the turbine is shut down to prevent structural overload.

Other wind turbine control objectives exist, which complement the already seen ones for the mentioned operating regions (during region two, energy capture is maximized, and in region three pitch control is used to limit power production to a certain value). These include, for example resonance avoidance, where the rotor speed is controlled so as to avoid a known resonant frequency and gust load alleviation, where the turbine settings are momentaneously changed so as to minimize damage from a spike in the wind speed, amongst others.

2.3 Wind Farm

Wind turbines are often sited together in wind farms due to a number of advantages, such as the available space, higher wind speeds, reduced deployment costs of turbines, reduced costs of the electricity grid and reduced operation and maintenance costs. In spite of these benefits, the grouping of turbines poses a number of complications that often significantly affect their performance. Such complications arise due to the presence of the wake. The wake refers to the altered wind flow downstream of the wind turbine and its properties depend on *time*—subject to the varying operational setting of a turbine and varying surrounding flow—*space*—the behaviour of the wake is different depending on how far downstream it is from the rotor—and *external variables*—such as temperature gradients, which influence the stability of the atmosphere and consequently the evolution of the wake.

A schematic representation of this interaction is illustrated in Fig. 2.4. Turbines are placed together, and the undisturbed flow will encounter the leading row of turbines first. The modified flow follows and directly impacts turbines placed further downstream.

The typical wake behaviour when travelling downstream of a wind turbine is represented in Fig. 2.5, and its most relevant characteristics are highlighted:

- *Reduced flow velocity*: natural consequence caused by the extraction of energy from the flow by the upstream turbine

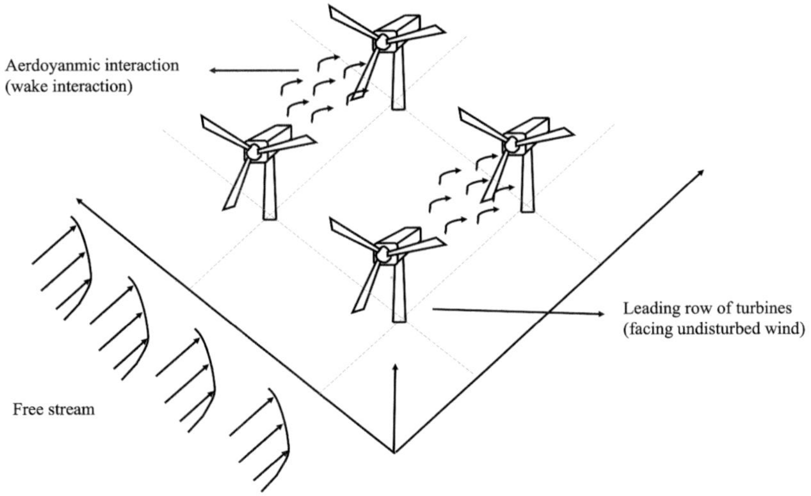

Fig. 2.4 Schematic representation of a simplified wind farm (a group of turbines, often placed offshore) and their aerodynamic interaction through the turbine's wake. On the left, the incoming undisturbed flow. On the right, in-between turbines, the wake represented through arrows flowing downstream

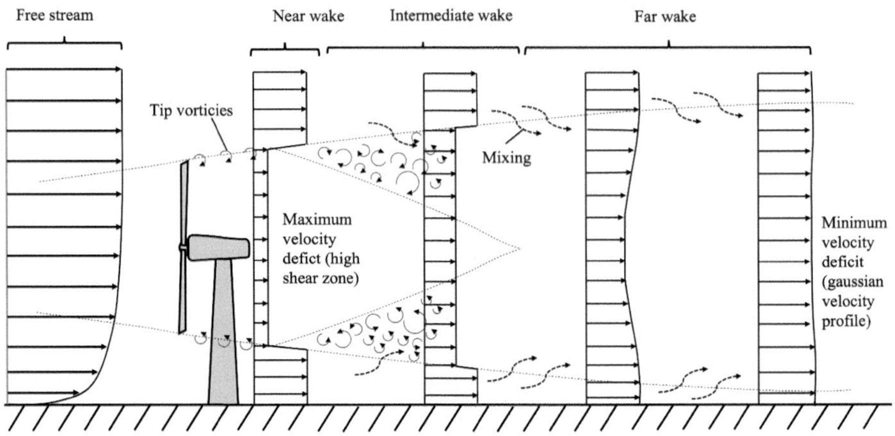

Fig. 2.5 Wind turbine wake behaviour development

- *Expansion of cross-sectional area*: explained by the law of mass conservation and the assumption that air density does not change (incompressibility)
- *Increased turbulence intensity*: caused by the obstruction of the flow by the turbine, and the resulting velocity gradients in the flow (shear)

The amount of wake interaction is also dependent on the distance between turbines, as the flow in the wake will recover to the surrounding flow conditions through convection and diffusion. Such processes are promoted by the turbulence, intensifying wake recovery by mixing of the flow in the wake and the surrounding stream, as depicted on the right side of Fig. 2.5. The increased turbulence and consequent wake recovery process is fuelled by the blades tip vortices which, initially stable, are transported downstream in the wake and eventually break down [4]. In addition, the wake rotates in the opposite direction to the rotor. Such characteristic is explained by Newton's third law of motion:[1] As the air passes through the rotor disk and exerts a torque, an equal and opposite torque is imposed by the rotor upon the downstream air. Other characteristic of the wake includes, for example, wake meandering, where the complete structure of the turbine wake moves on the vertical or horizontal direction.

The wake region closest to the turbine rotor is commonly referred to as *near wake*, and it may extend up to two rotor diameters[2] downstream [4]. The intermediate wake may extend up to three rotor diameters and the far wake region up until five rotor diameters, where the velocity follows a Guassian profile.

[1] Newton's third law states that for every action (force) in nature there is an equal and opposite reaction.

[2] Rotor diameter is a common unit to use as distance measurement in wind energy.

Different operating settings might induce other wake characteristics. As an example, in [5], for a yawed turbine (rotor is not oriented perpendicular with respect to the incoming wind), the wake is deflected, a phenomena where the complete wake is diverging in the perpendicular direction from the rotor centre. Moreover, the mentioned deflection increases with an increase in yaw angle and will present a kidney shape, as will be seen later on in Chap. 5.

2.4 Wind Farm Control

It is through the wake that an upwind turbine can influence the performance of downwind turbines. As a result of the downstream wind velocity deficit, downstream turbines will capture less power when compared to free stream conditions. Moreover, the increased turbulence in the wake can lead to increased loading, resulting in the wind turbine shorter lifetime.

The purpose of wind farm control is to go one step further than individual turbine control and take into account the interaction of the wake with the turbine to ensure a specific level of performance of the wind farm. This is achieved with a supervisory controller that acts on information provided by each individual turbine to reach a global objective. The objectives of wind farm control are similar to the ones of single turbine control: (1) maximize power production, (2) minimize structural degradation and (3) improve energy quality.

Wind farm control is a term which is often used when referring to manipulating the wake effect with the aim of increasing AEP or reducing loads; however it refers also to the control of the complete plant for other purposes. For example, due to restrictions from the grid operator, it may be necessary to reduce the active power production, often referred to as curtailment. Therefore, the term Active Wake Control (AWC) or Wind Farm Flow Control (WFFC) is often employed to refer only to the specific control feature which focuses on manipulating wake interaction. Nevertheless, for the sake of simplicity, the term wind farm control will only be used throughout the rest of this book to refer to manipulating the wake interaction.

Currently, most wind farms are operated using individually optimal wind turbine control settings, referred to as greedy control as a result of neglecting wake effects [6]. The assumption that wind farm performance can be increased by operating turbines in the farm at configurations different from their individual optimal control settings has become an active research topic since the 1990s.

Wind farm control is an active and growing field of research. Consensus of research findings, best practices and identification of open research questions are needed, as mentioned in [7], a recent expert elicitation on the field. The latter points towards reaching common standard definition of concepts, metrics and tools, as recommended by survey results and Topical Expert Meetings (TEMs) discussions.

Wind farm control focuses on plant level as opposed to turbine level control. One of its objectives is to increase Annual Energy Production (AEP) by decreasing the power losses due to the wake effect. Wind farm control makes use of three types of distinct technologies to achieve the aforementioned objective. The three are wake steering, axial induction control and wake mixing. The three differ on their inner workings, ability to achieve wind farm power gains and maturity given different levels of experimental validation. Figures 2.6 and 2.7 show the differences in terms of the modifications induced by the different flow control strategies in the wake's properties.

The three wind farm control technologies are the following [7]:

1. *Wake steering*: The wake is manipulated in such a way that its direction is changed so that it does not directly hit the downstream turbine rotor. This technique can be used to either increase power production, as the wake's velocity deficit at the downstream rotor is reduced, or reduce asymmetric loading introduced by partial wakes.

2. *Axial Induction Control*: Wind turbines are derated or uprated so that wake interactions and consequent impacts are modified. Derating entails operating

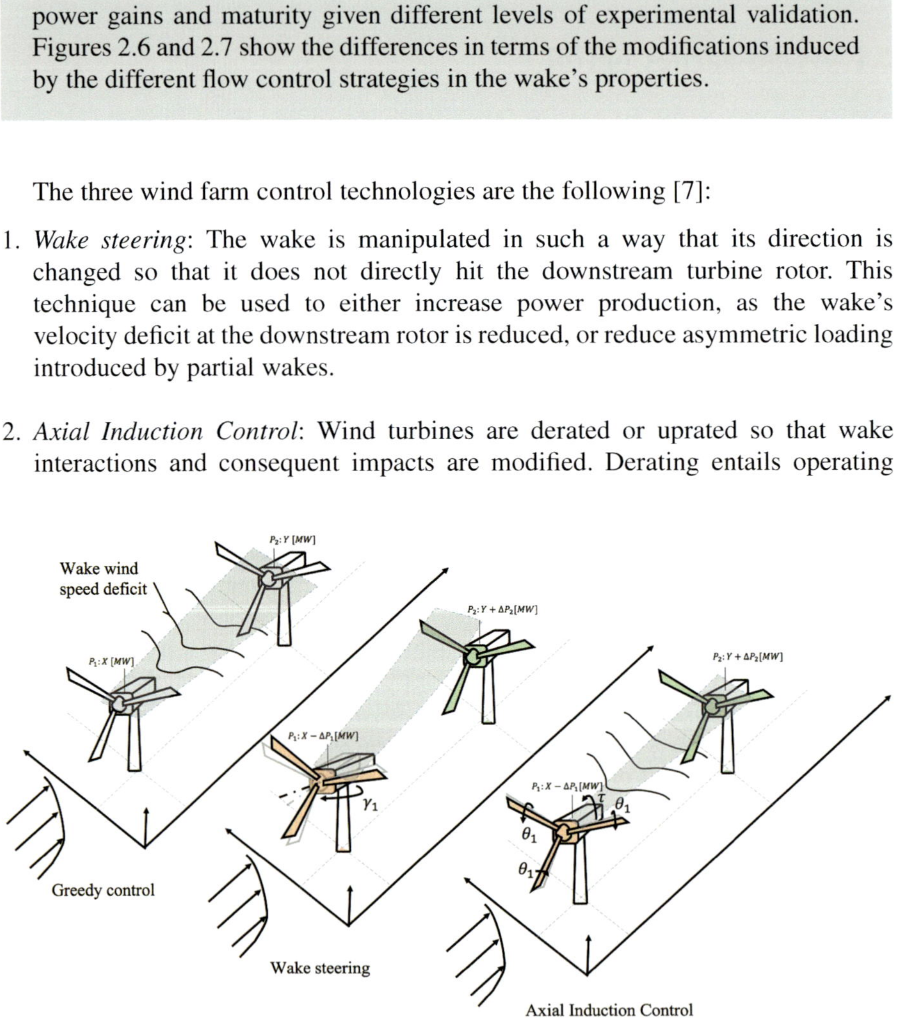

Fig. 2.6 Wind farm control strategies illustration. From left to right: greedy control, wake steering and Axial Induction Control. Greedy control refers to the operation of the turbines in the wind farm at their theoretical maximum. For wake steering and Axial Induction Control, the orange colour represents the sacrifice in terms of power lost in the upstream turbines and the green colour the theoretical gain at the downstream turbine

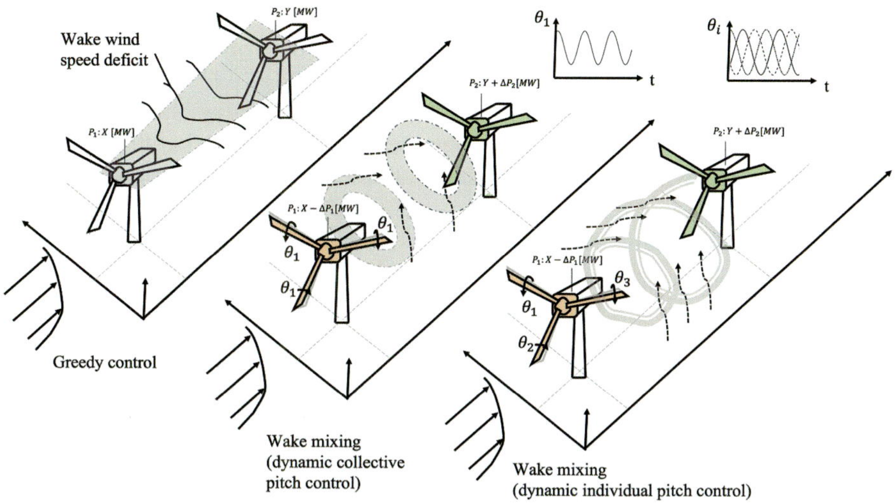

Fig. 2.7 Wind farm control strategies illustration. From left to right: greedy control, dynamic collective pitch control (also dubbed the pulse strategy) and dynamic individual pitch control (also referred to as the helix strategy). Dynamic collective pitch control generates pulsating structures. Dynamic individual pitch control generates a helicoidal wake that induces faster recovery

a wind turbine at a lower power than it could deliver (based on the wind conditions and the turbine's power curve). Upstream wind turbines can be derated to decrease structural loads (of both upstream and downstream turbines), and downstream turbines can be uprated to increase power production. Both uprating and derating are also the basis for providing grid services, such as active power control. It is often the case that the power grid needs to be stabilized and the wind farm needs to comply with the orders of the power grid operators. The power setpoints of turbines are then changed and turbines are down-regulated. This is achieved by modifying turbine's control inputs such as the blade pitch angle and/or generator torque.

3. *Wake mixing:* Upstream wind turbines are dynamically uprated and downrated on short time scales, thereby inducing additional wake recovery [7]. The goal is to reduce the velocity deficit further downstream, minimizing losses and increasing power production.

The three control strategies are schematically represented in Figs. 2.6 and 2.7. Figure 2.6 schematically illustrates wake steering and Axial Induction Control, whereas Fig. 2.7 depicts wake mixing strategies, such as dynamic pitch control (collective and individual), where the control inputs are varied at a certain frequency. These are further discussed in the following subsections.

Wind farm control is an active topic of interest around the world, which motivated the existence of an international co-operation on the topic in order to share information and growing novel research, the International Energy Agency

(IEA) Wind Technology Collaboration Programme (TCP) Task 44: Flow Farm Control. An outcome of this joint effort is the creation of a *wiki* page, i.e. a collaborative web page where research results, terminology and expert consensus are contained. At the moment this book was written, the creation and development of this page is an on-going effort. The web page can be accessed at the following location: https://ieawindtask44.tudelft.nl/. In addition, it is of relevance to mention that an interesting topic of research and part of the IEA TCP Task 44 relates to the evaluation of uncertainties surrounding the assessment of benefit from wind farm control technologies.

2.4.1 Wake Steering

Wake steering or Wake Redirection Control (WRC) is a strategy that attempts to change the direction of the wake so that the overlap of the wake with downstream rotors can be avoided. It can be achieved by yaw misalignment or Individual Pitch Control (IPC), although the benefits of the latter are more pronounced in terms of enhancing *wake mixing*. Another possibility for WRC is by tilting the rotor, meaning to lean the rotor forwards or backwards, although it is not a controllable feature in present-day wind turbines.

A simplified physical explanation is represented in Fig. 2.8. The thrust force T that the rotor exerts on the flow is shown along the axis of the rotor. When the wind inflow is at an angle to this direction due to misalignment, the thrust can be divided into its components T_x and T_y. The component T_x is parallel to the flow and slows the wind, while T_y is perpendicular and applies the force that causes wake redirection. Figure 2.8 depicts a schematic illustration of the wake being deflected behind the first turbine due to the component T_y.

This strategy has been proved to yield good results in simulations [8], wind tunnel experiments [9] and, more recently, field tests in wind farms [10–13]. Moreover, wake behaviour as a result of yaw actuation is an actively research topic in order

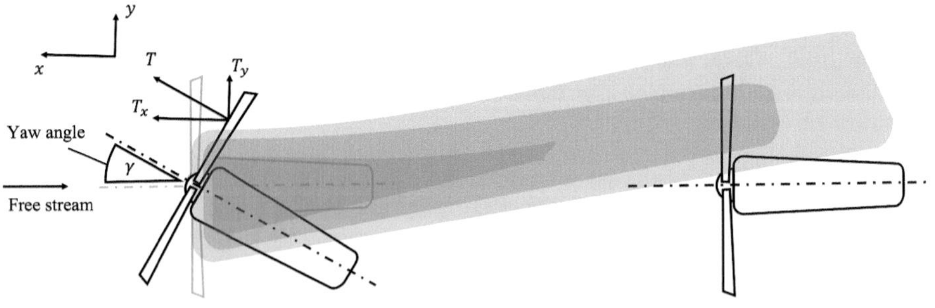

Fig. 2.8 Wake steering through yaw misalignment. The upstream turbine is purposely misaligned with relation to the incoming wind direction, creating a vertical component of the thrust force that redirects the wake. Overlapping with downstream turbine rotors is then avoided. Darker grey areas illustrate higher velocity deficits

to predict wake flows and provide valuable physical insights on the behaviour of turbine wakes. For a thorough discussion of the physical phenomena of the wake characteristic of a yawed wind turbine, the reader is referred to [5].

High fidelity simulations in Simulator for Offshore Wind Farm Applications (SOWFA),[3] a simulator of high fidelity, [8] have showed significant redirection effects coupled with reductions in loading. Wind tunnel tests have reproduced promising results as well. In [9], two scaled wind turbines with a rotor diameter of 1.9 m are placed apart, with the second one located at a distance of four rotor diameters downstream. It is then shown that power increases of up to 21% are achievable for a partial wake condition with a lateral offset of ± 0.45 rotor diameters.

Field tests in commercial wind farms have also been performed to better assess the benefits of wake steering. In [10], wake steering is demonstrated at an onshore six-turbine array, showing significant gains in power production for particular situations. The control scheme developed increased the power production for wind speeds near the site annual average between 7 and 13% and decreased variability by up to 72% for selected wind directions at night. Nevertheless, the net gain over annual production is reported to be insignificant when compared to baseline operation.

Additional field campaign results exist, with an approximate 14% increase in energy being measured on the downstream turbine for two closely spaced turbines in [11]. The combined upstream-downstream turbine pair in the latter article registered a 4% increase in energy production. Another example is [12], whose results indicate the ability of a wake steering controller to successfully increase power capture by amounts similar to those predicted from engineering models (engineering models and other types of wake models will be discussed in the next chapter).

Furthermore, wake steering is further demonstrated in a complex onshore wind farm in Italy [13], containing different turbine types and more complicated wake interaction. The measurements show increases of up to 35% for two turbine interactions and up to 16% for three turbine interactions. However, losses in power production are seen for various regions of wind directions.

In [14], dynamic strategies for yaw are also analysed, and it is found that dynamic yawing provides significant benefits over static yaw control in turbulent flow environments, whereas that is not the case for a uniform inflow profile, i.e. where the incoming wind conditions are variable in space-time.

One might think that due to the yaw misalignment, wake steering increases loads of the wind turbine. However, studies suggest otherwise. In [15], the authors conclude that the reduction in loads of the downstream turbine, as a result of the wake having been moved to the side, is more pronounced that the loads impact to the rotor misalignment of the upstream turbine. In addition, when taking into account the frequency distribution of wind speed and wind direction at a given site, the authors show that the lifetime fatigue loads[4] of the wind farm tend to decrease under wake redirection control by yaw misalignment.

[3] A more thorough explanation of SOWFA is provided in Sect. 4.1.1.

[4] Fatigue load is the load a material or component experiences by application of cyclical forces.

The assessment of turbine wake interaction, wake steering benefits and structural loads is still an active topic of research. The American WAKE experimeNt (AWAKEN) is a considerable effort in bridging the existing gaps in knowledge related to wind farm control and worth mentioning. It consists of a large-scale measurement campaign taking place in the Unites States mid-west. As stated in [16], *the overall objective of this project is to gather high fidelity observations of wind turbines and power plants operating in representative atmospheric conditions and to use the data to advance the understanding of wind plant physics.*

Wake steering is a Wind Farm Control strategy where the upstream turbine yaw angle is purposely misaligned. The wake is directed to one of the sides and avoids overlap with downstream turbine rotors. This is one of the most promising technologies, with simulation studies, wind tunnel experiments and recently test fields showing its potential.

The current state of the art for wake steering is usually based on a static optimization. In other words, the optimal yaw misalignments for each turbine that maximize the plant power production are computed offline for each wind speed and wind direction, based on an underlying wake model, which predict wind farm power production considering wake impingement on downstream turbines. This results in a look-up table (LUT), i.e. a set of optimized yaw angles for each turbine given certain wind conditions. This is illustrated in Fig. 2.9, where each turbine is misaligned with a pre-computed angle so as to redirect its wake.

Recent research has been focusing on the optimization of yaw angles under uncertainty as opposed to the static one. In [17], uncertainty in the yaw position is accounted in the procedure of optimizing the turbines yaw misalignment for wake redirection. It is seen that the optimized yaw misalignment setpoints are reduced, leading to a strategy which attempts to extract more energy from the upstream turbines. Furthermore, a case study with wind farm Princess Amalia showed that a robust optimized wind farm control scheme could lead to an annual average improvement of 0.2%.

In [18], an optimization problem under uncertainty is further evaluated, where uncertainties in the yaw offsets, wind direction, turbulence intensity and shear of the incoming wind are considered. Different Probability Density Functions (PDF) are fit onto the chosen variables based on existing studies, showing once again that the robust optimization leads to diminished yaw misalignments whilst simultaneously increasing in power for specific wind sectors, as compared to the deterministic conventional approach. The authors conclude with a futuristic note that the future of plant level control strategies will increasingly consider uncertainty.

The article [19] is another example of robust Active Wake Control (AWC), and it shows that plant controllers optimized under uncertainty are more beneficial than ones developed deterministically. In fact, it further shows that the deterministic

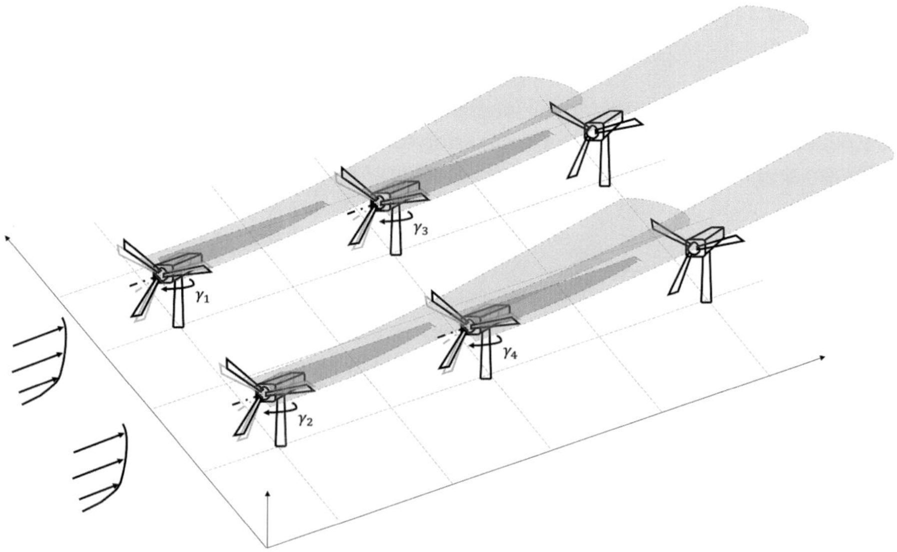

Fig. 2.9 Illustrative representation of wake steering state of the art approach in a real wind farm, where the yaw angles are optimized offline and stored in a look-up table

solution, when tested in simulations where the wind is modelled with a Gaussian distribution as opposed to a simplified mean value, may even lead to power losses.

In [20] a so-called dynamic robust AWC algorithm is developed. The variability of specific quantities such as the wind direction and wind velocity, amongst others, are modelled as stochastic processes by specifying the corresponding probability distribution functions.[5] The yaw misalignment setpoints are then optimized taking into account the uncertainties. Furthermore, as opposed to a static approach, a dynamic adaptation algorithm is developed, so that the yaw setpoints can take into account the spatio-temporal variability of the wind. The latter algorithm seeks to increase the plant's power production whilst minimizing the yaw duty of turbines under realistic operating conditions.

2.4.2 Axial Induction Control

The reasoning underpinning Axial Induction Control (AIC) is that by changing the axial induction—making use of the control variables available, blade pitch angle θ and generator torque τ—downstream turbines can generate more power. The

[5] Probability distribution functions are equivalent to Probability Density Functions, however defined for discrete variables.

control mechanisms influence the axial induction of the rotor and therefore also affect the velocity deficit in the wake the rotor generates. Axial Induction Control is worthwhile as long as reduced power production of the upwind turbine (the first of a series of turbines) is compensated for by the downwind turbines. A natural consequence is the reduction in thrust force, which leads to reduced turbulent wake mixing and thereby reduced wake recovery. There are thus two counteracting effects: increased velocity in the near wake but reduced recovery downstream. A simplified schematic representation for the classical axial induction control is provided in Figs. 2.10 and 2.11.

Despite initial promising results, recent studies with high fidelity simulation models and scaled wind tunnel experiments indicate that the power gain that can be achieved with steady-state induction control is limited to non-existing. In [21] it

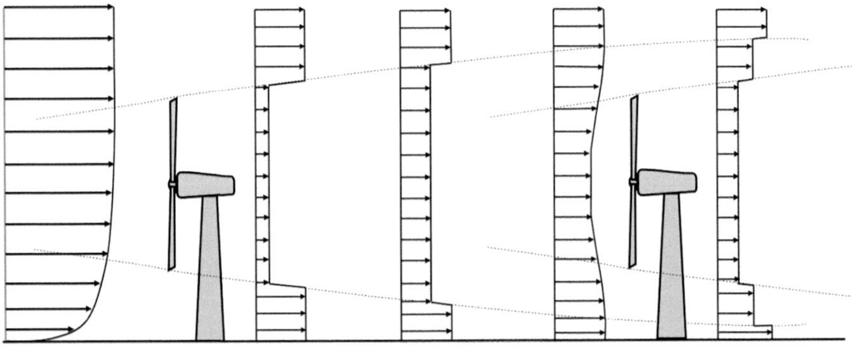

Fig. 2.10 Wake behaviour downstream of leading turbine and impact on downstream turbine using greedy control

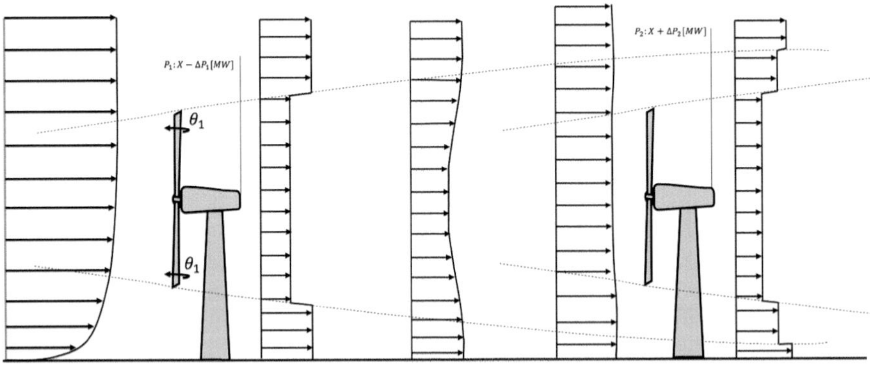

Fig. 2.11 Wake behaviour downstream of leading turbine when Axial Induction Control is employed. As first turbine curtails by pitching the blades, more energy is maintained in the wake. Velocity deficit in the wake is less pronounced as compared to the greedy case. Downstream turbine power production is higher due to increased local wind speed

is shown that the control settings that yield gain at the wind plant level are dependent on particular atmospheric conditions, the wind plant configuration and turbine characteristics. Simulations in SOWFA further suggest the existence of certain circumstances where this strategy becomes infeasible and that the power lost on the first turbine is not regained at the second turbine. Wind tunnel tests point again in the same direction: In [9], only very limited power gains are shown, showing a modest increase in total power of circa 1% when two turbines are considered.

2.4.3 Wake Mixing

Alternative approaches have emerged, which focus on time-varying control inputs which purposely influence the inherently dynamic nature of the wake. Recent studies, namely [22, 23], have shown that the so-called Dynamic Induction Control (DIC), where the induction factor is varied over time, improves power production in small- to medium-sized wind farms by generating a turbulent wind flow which enables enhanced wake recovery.

In the former article, power gains for wind farms with optimal coordinated control in the order of 8–21% compared with a greedy controlled case are found. In the latter article, these results are explored in more depth, where the simpler approach of varying the induction sinusoidally is taken. A best parameter for the sinusoidal variations is found, corresponding to an amplitude A^* of 1.5 and a non-dimensional frequency St^*[6] of 0.25 , proving to be robust to both wind turbine spacing and turbulence intensity. Wind tunnels experiments with DIC detailed in [24] prove the potential of such control strategy in real-world experiments, showing a potential increase in power production up to 4% in a three scaled wind turbine set-up.

The physical mechanisms leading to wake break up using DIC are still an open area for research. As mentioned in [25], *whether it is possible to control turbulence at a larger boundary layer scale for improved mixing and energy extraction also remains an interesting fundamental research question.* The work in [26] aimed at shortening the knowledge gap by investigating the wake properties induced by DIC through a wind tunnel experiment. In the latter, the authors investigate the wake of a small-scale wind turbine with a diameter of 58 m operating with DIC. The wake was measured using three dimensional Particle Image Velocimetry (PIV), a technique commonly used in wind tunnel experiments for these purposes. Decreases in the wake velocity and increases in turbulence intensity towards the centre of the wake were observed when compared to the baseline experiment, evidencing the enhanced wake recovery. An additional indication of the increased recovery was the increase in available power downstream. The authors also noted the simultaneous interaction

[6] The adimensionalized frequency is quantified by the Strouhal number, defined as $St = fD/U_\infty$ for a given frequency f, rotor diameter D and inflow velocity U_∞.

of three successive vortexes, hypothesizing that the new observed mechanisms of vortex breakdown are responsible for the enhanced wake recovery than continuous downstream and expands over a large area, emphasizing the need of additional research for a proper understanding of it.

The dynamical concepts used in the context of AIC have also been leveraged with Individual Pitch Control (IPC). Such approach is labelled the Dynamic Individual Pitch Control (DIPC), where the goal is to dynamically vary the tilt and yaw moments such that the wake of the turbine is manipulated in the vertical and/or horizontal directions, over time. Individual Pitch Control, in its simplest form, creates an uneven distribution of thrust forces on the rotor blades over the course of a rotation and uneven blade torques over the course of a rotor rotation as well. The former will not cause significant redirection of the flow behind the turbine, as the thrust reaction forces on the flow are directed parallel to the inflow direction. The latter, however, will cause the flow to be redirected and the wake structure to be skewed, as the turbine applies a net force on the flow perpendicular to the thrust direction. These imbalances cause moments at the centre of the rotor, the tilt M_{tilt} and yaw M_{yaw} moments. The completely novel concept published in 2019 is described in [27]. The resulting wake propagates through space in a helical fashion, as schematically represented in the right most illustration in Fig. 2.7, and the flow control strategy is referred to as the *helix strategy*. The article shows it can be effective as a wind farm control strategy, improving wind energy capture and decreasing power and thrust variations due to enhanced wake mixing.

This chapter started by introducing the main physical mechanisms related with the extraction of wind energy by a horizontal axis wind turbine. It then introduced the control variables available to maximize the extraction of energy, giving rise to the different operating regions. However, it was explained that the maximization of energy at the turbine level, often referred to as greedy control, does not necessarily lead to the maximum extraction of energy by the plant due to the wake effect. Nevertheless, by making use of the turbine level control variables, the losses can be partly overcome. Using the latter at the farm level is denominated as wind farm control and serves the same objectives as turbine control, such as active power regulation, energy extraction maximization, amongst others. The maximization of energy production through direct manipulation of the wake effect is often referred to wind farm flow control. Different wake control technologies exist, being distinguished between wake redirection control, axial induction control and wake mixing. All of these differ in their inner workings and level of maturity, with wake redirection control being already tested in wind farms and wake mixing being in its infancy. Current research has been focusing on validating these technologies in the field and through wind tunnel experiments. A deeper understanding of the physical mechanisms explaining the gains related with the more recent technologies is still necessary, and modal decomposition may assist in this. In the next chapter the core algorithm explored in this book, a variant of the dynamic mode decomposition, is introduced.

References

1. J.F. Manwell, G. McGowan, A.L. Rogers, *Wind Energy Explained*, 2nd edn. (Wiley, Hoboken, 2008)
2. N. Jenkins et al., *Wind Energy Handbook*. English, 3rd edn. (Wiley, Hoboken, 2001)
3. F.D. Bianchi, D.B. Hernán, R.J. Mantz, *Wind Turbine Control Systems : Principles, Modelling and Gain Scheduling Design*, 1st edn. (Springer, Berlin, 2007)
4. I. Neunaber et al., Distinct turbulent regions in the wake of a wind turbine and their inflow-dependent locations: the creation of a wake map. Energies **13**(20), 5392 (2020). ISSN: 1996-1073. https://www.mdpi.com/1996-1073/13/20/5392.
5. M. Bastankhah, F. Porteé-Agel, Experimental and theoretical study of wind turbine wakes in yawed conditions. J. Fluid Mech. **806**, 506–541 (2016)
6. S. Boersma et al., A tutorial on control-oriented modeling and control of wind farms, in: *2017 American Control Conference (ACC)* (2017), pp. 1–18
7. J.W. van Wingerden et al., Expert elicitation on wind farm control. J. Phys. Conf. Ser. **1618**, 022025 (2020)
8. P. Fleming et al., Evaluating techniques for redirecting turbine wake using SOWFA. Renew. Energy **70**, 211–218 (2014)
9. F. Campagnolo et al., Wind tunnel testing of wake control strategies, in *2016 American Control Conference (ACC)* (2016), pp. 513–518
10. M.F. Howland, S.K. Lele, J.O. Dabiri, Wind farm power optimization through wake steering. Proc. Natl. Acad. Sci. **116**(29), 14495–14500 (2019). ISSN: 0027-8424
11. P. Fleming et al., Initial results from a field campaign of wake steering applied at a commercial wind farm—Part 1. Wind Energy Sci. **4**(2), 273–285 (2019)
12. P. Fleming et al., Field test of wake steering at an offshore wind farm. Wind Energy Sci. Discuss. **2**, 1–17 (2017)
13. B.M. Doekemeijer et al., Field experiment for open-loop yaw-based wake steering at a commercial onshore wind farm in Italy. Wind Energy Sci. Discuss. **2020**, 1–22 (2020)
14. W. Munters, J. Meyers, Dynamic strategies for yaw and induction control of wind farms based on large-eddy simulation and optimization. English. Energies **11**(1), 177 (2018). ISSN: 1996-1073
15. S. Kanev, E. Bot, J. Giles, Wind farm loads under wake redirection control. Energies **13**(16), 4088 (2020). ISSN: 1996-1073. https://doi.org/10.3390/en13164088. https://www.mdpi.com/1996-1073/13/16/4088
16. P. Moriarty et al., American WAKE experiment (AWAKEN). Technical Report, National Renewable Energy Laboratory (2020)
17. J. Quick et al., Optimization under uncertainty for wake steering strategies. J. Phys. Conf. Ser. **854**, 012036 (2017). https://doi.org/10.1088/1742-6596/854/1/012036
18. J. Quick et al., Wake steering optimization under uncertainty. Wind Energy Sci. **5**(1), 413–426 (2020). https://doi.org/10.5194/wes-5-413-2020. https://wes.copernicus.org/articles/5/413/2020/
19. A. Rott et al., Robust active wake control in consideration of wind direction variability and uncertainty. Wind Energy **3**, 869–882 (2018). https://doi.org/10.5194/wes-3-869-2018
20. S. Kanev, E. Bot, Dynamic robust active wake control. Wind Energy Sci. Discuss. **2021**, 1–30 (2021). https://doi.org/10.5194/wes-2021-71. https://wes.copernicus.org/preprints/wes-2021-71/
21. J. Annoni et al., Analysis of axial-induction-based wind plant control using an engineering and a high-order wind plant model. Wind Energy 19, 1135–1150 (2015)
22. W. Munters, J. Meyers, An optimal control framework for dynamic induction control of wind farms and their interaction with the atmospheric boundary layer. English. Philosoph. Trans. R. Soc. London, Ser. A Math. Phys. Eng. Sci. **375**(2091), 20160100 (2017). ISSN: 1364-503X

23. W. Munters, J. Meyers, Towards practical dynamic induction control of wind farms: analysis of optimally controlled wind-farm boundary layers and sinusoidal induction control of first-row turbines. English. Wind Energy Sci. **3**(1), 409–425 (2018). ISSN: 2366-7443. https://doi.org/10.5194/wes-3-409-2018
24. Y.A. Frederik et al., Periodic dynamic induction control of wind farms: proving the potential in simulations and wind tunnel experiments. Wind Energy Sci. Discuss. **5**, 1–18 (2019)
25. J. Meyers et al., Wind farm flow control: prospects and challenges. Wind Energy Sci. **7**(6), 2271–2306 (2022). https://doi.org/10.5194/wes-7-2271-2022. URL: https://wes.copernicus.org/articles/7/2271/2022/.
26. D. van der Hoek et al., Experimental analysis of the effect of dynamic induction control on a wind turbine wake. Wind Energy Sci. **7**(3), 1305–1320 (2022). https://doi.org/10.5194/wes-7-1305-2022. https://wes.copernicus.org/articles/7/1305/2022/
27. J.A. Frederik et al., The helix approach: using dynamic individual pitch control to enhance wake mixing in wind farms. Wind Energy **23**(8), 1739–1751 (2020)

Chapter 3
Data-Driven Modelling of Wind Farm Control Strategies

This chapter introduces the task of modelling a wind farm. Different turbine and flow models are presented. Data-driven techniques previously used within the context of farm modelling are considered. The core technique used in this book, Dynamic Mode Decomposition, is discussed in great detail. Its variants suited for the purpose of modelling systems where inputs and outputs are considered, suited for the purposes of control, are presented. In addition, variants of these techniques motivated by mathematical concepts from non-linear dynamic systems theory are discussed. This chapter focuses heavily on the data-driven techniques and presents them from a mathematical and algorithmic point of view.

3.1 Wind Farm Modelling

The task of modelling, broadly speaking, focuses on describing the relationships between variables of a given system. The system represents an object of interest, in which variables of different kinds interact and produce observed signals. The system of interest which will be the main focus in later chapters of this book is a small wind farm, more specifically, a set of two turbines which interact via the wake. The relevant observed signals are referred as *outputs*, and the external signals manipulated by the observer which serve as a stimulus to the system are called *inputs*. The role of the model is then to incorporate these relations, so the user may perform simulations and predictions.

> The study of wind farm control algorithms requires an accurate description of the wake behaviour and its interaction with wind turbines, achieved through

(continued)

N. Cassamo, J.-W. van Wingerden, *Data-driven Modelling of Wind Farm Flow Control Strategies*, https://doi.org/10.1007/978-3-031-84116-3_3

the usage of wind farm models. These are comprised of wind turbine and flow models. The turbine model predicts the interaction between the turbine structure and the flow field, information that can be used to estimate loading on the turbine. The flow field refers to the flow properties in a wind farm. These two are coupled, with the turbine model requiring information from the flow field to perform calculations and vice versa [1].

3.1.1 Turbine and Flow Models

Although wind farm models may differ, three main components can always be distinguished, as described in [2]:

- *A model describing the flow characteristics in the wind field around the turbines*: Various models of different complexities offer results which naturally vary in accuracy. Models can be segmented in high fidelity, medium fidelity and low fidelity. As it follows from the category name, higher fidelity models provide the most accurate results. However, the computation cost, i.e. the effort required by the computer to solve the necessary underlying mathematical equations, is also the highest, where simulations can take several days. The opposite holds for the lower fidelity models. Higher fidelity models rely on solving the equations that describe the motion of a fluid in two or three dimensions. These equations are usually solved resorting to Computational Fluid Dynamics (CFD), a branch of fluid dynamics specializing in solving this type of problem.
  On the other side of the spectrum lie the lower fidelity models that are parametric in nature. These are simplified analytical formulations, where mathematical expressions describing key properties of the fluid are used. The expressions are a function of a limited number of input parameters. Such models are implemented, for example, in FLOw Redirection and Induction in Steady State (FLORIS) [3], a tool comprising wind plant models and optimization routines for different purposes (wind farm control, optimization of the farm layout). FLORIS implements models such as the Jensen [4], the Gaussian Curl Hybrid model, based on the works of [5–7], the TurbOPark model [8] (also available at https://github.com/OrstedRD/TurbOPark) and the Cumulative Curl model [9]. For more information on the different types of models and their distinct characteristics, the reader is referred to [1] or the mentioned articles. Models can differ in the underlying physical assumptions used, such as, for example, how different wakes are combined or the underlying turbulence model employed. Engineering wake models are comprised of a different number of parameters that govern the characteristics of the wake, such as the rate of growth or expansion.
  Validating different wake models to evaluate their suitability to predict power performance of wind farms is an active topic of research. In the recent work

of [10], the Gaussian Curl Hybrid Model is compared to historical data from three wind farms: Anholt, Offshore Wind farm Egmond aan Zee (OWEZ) and Westermost Rough. The three wind farms are placed offshore, however differ in the number of turbines (from 35 to 111), turbine spacing (from 4.9 to 7.2 rotor diameters) and type of turbine (from 3 to 6 MW, differing from 90 to 154 in rotor diameter), being an interesting exercise to evaluate the suitability of the widely used Gaussian model to a variety of parks. The authors show that for the majority of wind directions an excellent agreement between model predictions and historical data exists. There are however discrepancies, which are hypothesized to be due to two reasons. One is thought to be related with complex inflow conditions, such as heterogeneity (i.e. blockage effects, coastal effects and other effects from neighbouring turbines). The other is deep array effects, related with the interacting and overlapping of multiple wakes far often found downstream in a wind farm. This is specially observed for the turbine located in the centre of Anholt. The Cumulative Curl model developed in [9] aims to improve on the aforementioned discrepancies by, for example, replacing the near-wake model used in the Gaussian one and replacing the wake deficit and wake superposition sub-models by the cumulative model of wind turbine wakes in wind farms proposed in other research. It is shown using data from high fidelity simulations and from physical wind plants historical data that the Cumulative Curl model benefits from improved accuracy when compared to the Gaussian one in situations with large wake losses and for specific situations where inter-turbine distances are considerable.
Another recent work of interest is [11], where the authors validate the Park and TurbOPark model to historical data from 19 offshore wind farms. This fits within the larger task of developing a systematic framework for validation and uncertainty quantification of wind farm models. The authors find good agreement between the model and observed data and conclude that the model has a very low bias, which does not exhibit any particular trend based on the wind plant configuration. In addition, the authors further recommend using a specific value for the parameter governing wake growth throughout the farm, indicating that it is suited for offshore applications, and however, its validity must be tested for onshore.

- *A model of the forces of the wind turbine on the flow and the reactive forces of the flow on the wind turbine*: Two models usually used are the Actuator Disk Model (ADM)—the rotor is modelled as a disk of distributed forces exerted on the flow—and the Actuator Line Model (ALM)—a more detailed representation of the rotor where each blade is represented individually in the flow as a distribution of forces along a rotating line. Using the indicated models has significant advantages in terms of computational cost, as using an exact representation of the blade bodies becomes too costly for simulating wind plants.

- *A model describing the effect of the flow forces on the performance of the wind turbine system*: Estimations on power capture by wind turbines are contemplated in the ADM and ALM [1]. A more elaborate turbine model is the FAST, developed by the National Renewable Energy Laboratory (NREL). It contains the ALM and calculates the loads, power and rotor speed of each turbine.

3.1.2 Governing Dynamics

The motion of a fluid—and thus the wake—is governed by three basic differential equations: the continuity equation, the momentum equation and the energy equation (which is omitted for simplicity). The complete derivation can be seen in [12]:

$$\text{Continuity:}\ \frac{\partial \rho}{\partial t} + \nabla \cdot (\rho \boldsymbol{V}) = 0 \tag{3.1}$$

where $\cdot\nabla$ is the vector divergence operator,[1] ρ the fluid's density and $\boldsymbol{V}$ the velocity vector. The equation in (3.1) expresses the conservation of mass (within a certain volume).

$$\text{Momentum:}\ \rho \frac{d\boldsymbol{V}}{dt} = \rho \boldsymbol{g} - \nabla p + \nabla \cdot \boldsymbol{\tau}_{ij} \tag{3.2}$$

where ∇ is the vector gradient operator, $\boldsymbol{g}$ is the gravitational acceleration vector, p is the hydrostatic pressure and $\boldsymbol{\tau}_{ij}$ the viscous stress tensor. The equation in (3.2) expresses the conservation of momentum. For an incompressible Newtonian[2] fluid, it simplifies to the famous Navier Stokes (NS) equations, which, written in full, illustrate the inherent mathematical difficulties:

$$\begin{aligned} \rho g_x - \frac{\partial p}{\partial x} + \mu \left(\frac{\partial^2 u}{\partial x^2} + \frac{\partial^2 u}{\partial y^2} + \frac{\partial^2 u}{\partial z^2} \right) &= \rho \frac{du}{dt} \\ \rho g_y - \frac{\partial p}{\partial y} + \mu \left(\frac{\partial^2 v}{\partial x^2} + \frac{\partial^2 v}{\partial y^2} + \frac{\partial^2 v}{\partial z^2} \right) &= \rho \frac{dv}{dt} \\ \rho g_z - \frac{\partial p}{\partial z} + \mu \left(\frac{\partial^2 w}{\partial x^2} + \frac{\partial^2 w}{\partial y^2} + \frac{\partial^2 w}{\partial z^2} \right) &= \rho \frac{dw}{dt} \end{aligned} \tag{3.3}$$

where μ is the viscosity coefficient and (u, v, w) the velocity components in the cartesian coordinate system (x, y, z). The triplet of equations along with the incompressible form of the continuity equation (3.1) forms four equations in the four unknowns p, u, v and w. The NS equations are second-order Non-linear Partial Differential Equations (NLPDE) and have only a limited number of known analytical solutions.

[1] The divergence and gradient operators are detailed in Annex A.2.

[2] A Newtonian fluid is defined as one with constant viscosity due to the linear relation between applied shear and resulting strain rate.

The inherent mathematical complexity of the Navier Stokes equations, along with the large dimensions of the wind farm, is the driving force behind the using of other methodologies, such as data-driven and equation free techniques, for the purposes of modelling wind farms.

The flow may exhibit smooth and steady (laminar) or fluctuating and agitated behaviour (turbulent). Turbulence is characterized by the rapid and random fluctuations in the flow. Every pressure and velocity terms in Eqs. (3.1) and (3.2) is a rapidly varying function of time and space, leading to the unsteady NS equations. The dynamical behaviour of a wind turbine wake is governed by the 3D unsteady NS equations. These can be rewritten in terms of time-averaged turbulent variables, leading to the famous Reynolds Average Navier Stokes (RANS) equations. The velocity components are separated into their mean and fluctuating parts: $u = \bar{u} + u'$, $v = \bar{v} + v'$ and $w = \bar{w} + w'$. For x, often the most relevant direction aligned with the flow propagation, the equation takes the form (for a thorough derivation, please refer to [12]):

$$\rho \frac{d\bar{u}}{dt} = -\frac{\partial \bar{p}}{\partial x} + \rho g_x + \frac{\partial}{\partial x}\left(\mu \frac{\partial \bar{u}}{\partial x} - \rho \overline{u'^2}\right) + \frac{\partial}{\partial y}\left(\mu \frac{\partial \bar{u}}{\partial y} - \rho \overline{u'v'}\right)$$
$$+ \frac{\partial}{\partial z}\left(\mu \frac{\partial \bar{u}}{\partial z} - \rho \overline{u'^2}\right) \tag{3.4}$$

where the terms $-\rho \overline{u'^2}$, $-\rho \overline{u'v'}$ and $-\rho \overline{u'w'}$ are called *turbulent stresses*. The several terms mentioned are of interest to test various modelling approaches in Chap. 5.

3.2 Data-Driven Modelling of Wind Farms

Data-driven techniques facilitate the modelling of systems of high complexity, such as wind farms. The starting point is data from the systems, as opposed to the governing equations. Furthermore, these algorithms usually apply dimensionality reduction techniques, and the final model represents the dynamics of the system in a low dimensional space. The most well-known algorithms are Proper Orthogonal Decomposition and Dynamic Mode Decomposition. This book focuses on the latter.

A typical high fidelity wind farm simulation might contain 10^6 or more states—points where the equations are numerically solved and the flow field information retrieved. However, building models with such complexity might prove computationally unfeasible. An alternate solution is to perform model order reduction techniques,[3] thus diminishing the model's complexity whilst still benefiting from the accuracy of high fidelity simulations. Furthermore, the flow fields obtained with the lower order model with that of a high fidelity flow solver can be compared.

Examples of such techniques are Proper Orthogonal Decomposition (POD) and Dynamic Mode Decomposition (DMD). These provide a possible route to real-time implementation of optimal controllers, as mentioned in [13], considering that simulation-based optimal controllers using the complete flow field information have high computational costs and are hence not suitable for designing practical cooperative wind farm controllers.

In addition, Reduced Order Models (ROMs)—the final model resulting from applying model order reduction techniques—can be pure data driven, meaning no prior knowledge of the governing equations is necessary, which is of significant advantage given the complexity of the unsteady NS equations. Nevertheless, it should be noted that these techniques rely on specific operating conditions, meaning they provide linear models for a specific operating point and are only valid within small deviations from this point [1]. In spite of that, in [14], it is shown that reduced order Linear Parameter Varying (LPV) models can be constructed to overcome the stated limitation. Data from different operating conditions is used, and different models at these conditions are computed and stitched together.

3.2.1 Proper Orthogonal Decomposition

Proper Orthogonal Decomposition (POD) is used to analyse experimental data with a view of extracting dominant features and trends—in particular, the so-called *coherent structures*. A coherent structure is a type of repetitive pattern seen in a turbulent fluid flow, maintaining the same size or shape through time, in spite of the whole flow being unsteady and not repeating itself in detail. An example of these structures are the wakes and a pair of counter-rotating vortices formed on the surface of flowing water behind obstacles, long ago sketched by Leonardo Da Vinci in his notebooks [15]. Therefore, POD, in the context of turbulence and other complex spatio-temporal fields, will typically extract patterns in space and time.

At the heart of the POD is the Singular Value Decomposition (SVD) of the experimental data matrix $\mathbf{X}$ (the concepts of data and snapshot matrices, relevant to data-driven methods, are further elaborated in Sect. 3.3). The SVD is a mathematical

[3] Model order reducing techniques refer to a set of mathematical techniques that reduce the number of variables in a model which are necessary to reproduce the behaviour of the system, preserving the important dynamic of the original model.

operation which consists of a matrix factorization,[4] and applying it to $\mathbf{X}$ yields

$$\mathbf{X} = \mathbf{U}\boldsymbol{\Sigma}\mathbf{V}^* \tag{3.5}$$

Proper Orthogonal Decomposition separates the space and time correlations into the matrices $\mathbf{U}$ and $\mathbf{V}$, respectively. The matrix $\mathbf{U}$ only contains the spatial correlations in the data, whereas all temporal information is found in $\mathbf{V}$. The columns of $\mathbf{U}$ are called POD modes, which provide an orthogonal basis[5] of spatially correlated modes. Its properties suggest that it is the preferred basis to use in various applications [15]. The singular values $\boldsymbol{\Sigma}$ represent the relative energy in each mode. In this way, this method determines the most energetic structures.

Nevertheless, drawbacks from employing this technique arise and are pointed out in [16]: The energy may not in all circumstances be the correct measure to rank the flow structures. This shortcoming has been widely recognized and serves as an explanation for the existence of dynamically highly relevant but zero energy modes. Dynamic Mode Decomposition (DMD) emerges as a decomposition method based on data which is able to yield fluid structures that accurately describe the motion of the flow, overcoming the shortcoming of POD. This technique produces structures of a linear tangent approximation to the underlying flow, thus describing fluid elements that express the dominant dynamic behaviour captured in the data, as the author in [16] emphasizes.

Applications of POD to the modelling of wind turbine wakes exist in the literature. This book focuses on the application of DMD (and variant algorithms), and for more information on the application of POD, the following sources can be consulted: [17–20]. It is of interest, nevertheless, to note some insights from these articles.

The first is that many of the authors find POD useful to have a better understanding of the complex phenomena being studied. For example, the authors in [19] mention that the wake dynamics relevant to a downstream turbine, such as the energy flux through a disk, can be captured using the three most energetic modes found. The authors further emphasize the point that simplified dynamic wake models based on superposition of modes should be possible and that these could play an important role for the layout optimization and controlling of wind farms. In [17] the so-called Double Proper Orthogonal Decomposition (DPOD) is applied, where POD is first used to decompose and analyse the different structures in the wind turbine wake and then applied again to a set of modes from each separate POD basis to evaluate the existing sub-modal structures.

The second is that some of the already mentioned shortcomings of POD are once more raised. For example, the author in [19] mentions that including modes in the usual energetic order is not generally the best choice, pointing towards the same

[4] More details on the SVD in Annex A.1.

[5] An orthogonal basis is a set of vectors that are perpendicular (orthogonal) to each other, thus being linearly independent form each other.

conclusions as in [16]. In addition, the author further acknowledges the inherent difficulty in building a ROM based on the extracted POD modes.

3.2.2 Dynamic Mode Decomposition

Dynamic Mode Decomposition is the technique this book focuses upon. It allows to compute a simplified ROM of a high complex system, and similarly to POD, to represent the dynamics of interest—in this case, the wind turbine wake—as a linear superposition of modes. Dynamic Mode Decomposition makes use of the abundance of data retrieved from available samples of the system of interest to reverse engineer the underlying equations, without having to explicitly use them, as the author in [21] cleverly summarizes. This is applicable and highly valuable to many modern systems, such as epidemiological systems or financial markets which are characterized as being high dimensional, non-linear systems which exhibit rich multi-scale phenomena in space and time. The underlying goal of DMD, regardless of the system, is always the same: to extract meaningful spatio-temporal patterns that dominate dynamic activity of the system being studied.

The unprecedented availability of high fidelity measurements from numerical simulations and experimental data, the abundance of computational resources and capacity for digital storage and the *big data regime*[6] have provided the opportunity and created the incentive for the rapid development of these algorithms and their applications in different and relevant domains of science and engineering.

Dynamic Mode Decomposition is a technique that decomposes a system in terms of its most (dynamically) relevant spatio-temporal structures. The starting point is data from the system. The DMD algorithm can be used to gain insight into the underlying behaviour of the system by studying the spatio-temporal patterns which dictate the system's behaviour, build models of reduced dimensions of the system of interest useful for simulations and forecasting as these are more efficient and design controllers.

The method originated in the fluid dynamics community as a means to decompose complex flows into a simple representation, and it was first proposed by Schmid and Sesterhenn [16] and later published in [23]. In the latter, DMD is validated and applied for three different flow cases—the first dataset is generated numerically in a simulation and the last two based on data extracted from experimental flow measurements. As a result, Peter Schmid, author of the articles, proves DMD as a

[6] The big data regime refers to the gargantuan bulk of data which cannot be dealt with by traditional data-handling techniques, weather in terms of volume, velocity, or a variety of data [22].

robust and reliable algorithm to extract spatio-temporal coherent structures from the data.

Dynamic Mode Decomposition has been applied to a wide variety of flow geometries, and it has proven its value as a new quantitative flow analysis tool when compared to other methods. It is shown in [24] whereas POD concentrates on the more energetic structures of the flow, DMD isolates the less energetic but more unstable axial vorticity patterns close to the outer wall of the cylinder in the experiment, which may be dynamically more relevant.

Since then, different variations of DMD have surged. In [25] DMD is generalized to a larger class of datasets, including nonsequential time series. Furthermore, DMD has been leveraged in other domains of science. In the context of epidemiology [26], DMD allows for epidemiological interpretation of large-scale dynamic patterns such as how a pattern of disease infection spreads across a spatial domain. The insights uncovered by DMD are of the highest interest for planning annual resource allocation of vaccines and improving other decision-making processes. Within the context of neuroscience [27], DMD is used to extract coherent patterns in large-scale neural recordings to decipher structure's oscillatory behaviour generated in the brain during sleep, thus helping to tackle a major challenge in the field.

Dynamic Mode Decomposition has numerous uses. Using the same categorization of DMD's applications as in [21], these are:

1. *Diagnostics:* Extracting key spatio-temporal features of high dimensional systems, allowing for physically interpretable results in terms of spatial structures and their associated temporal responses. The user can then increase his understanding of the system being studied, which is of added value in light of the high degree of complexity associated in these systems.

2. *State estimation and future state prediction:* Anticipating the behaviour of the system where no measurements were made. This more sophisticated use of the DMD algorithm can be achieved by using the dominant spatio-temporal identified structures to construct dynamical models of the underlying processes observed.

3. *Control:* Enabling control strategies is the ultimate and most challenging goal of the DMD algorithm. The intrinsic difficulty is understandable as the final product of DMD—a linear dynamical model—is used to predict the future of a non-linear dynamical system. It is reasonably expectable that perhaps only a limited short time window in the future where the two models will actually agree exists.

The following chapters focus on the first two applications of DMD, focusing on how to capitalize on this algorithm and its variants to understand the underlying dynamics of two wind farm control strategies and the process of building a model with the most relevant dynamic spatio-temporal structures.

One of the first instances in the literature, dating from 2015, is [28], where a DMD model is embedded in a Kalman Filter[7] and shows that the entire wake flow can be predicted with a global accuracy of about 4%. The application, however, is only for a single turbine. A two turbine array setting appears then in [29], where the findings indicate that a ROM can be obtained and used as a predictive model for a similar dataset, where forced input is now considered.

Furthermore, in [14], a Linear Quadratic Gaussian (LQG)[8] based on a ROM is designed in order to minimize velocity fluctuations at a specified downstream location, and it is shown that the controller is able to reduce significantly fluctuations by varying the thrust force. In [30], a Model Predictive Controller (MPC)[9] is designed based on a ROM computed using a variant of the DMD algorithm. The MPC is then implemented in SOWFA so that two turbines can produce, together, a pre-defined amount of power. It is seen that the two turbines can produce the desired power by cooperating and taking the wake interaction into account. The controller uses information related to the future setpoint of increased power production and the upstream turbine yaws in advance, thus considering the time the deflected wake takes to travel and the impact in the power production of the downstream turbine.

In [31] another example of using a DMD related algorithm is shown. Although the focus is not in modelling the wind turbine wake, it is of relevance in the context of wind energy. In this chapter, the authors develop a variant of the DMD algorithm in order to work with streams of data from the supervisory Control and Data Acquisition System (SCADA) by updating the decomposition as new information is obtained. The methodology is dubbed streaming Dynamic Mode Decomposition (sDMD). The method is applied to data from offshore wind farm Lillgrund, and the authors show that the algorithm is able to outperform others such as persistence and moving average power forecasting for time windows in the scale of minutes. However, it is also acknowledged that for shorter time windows, such as thirty seconds, improvements with relation to other methods are not as obvious. The explanation provided is that relevant information on these time scales is lost due to the model reduction procedure.

The following subsection introduces the core DMD algorithm.

[7] Estimator that reconstructs the state of a dynamical system from measurements of a time series of the sensor outputs and actuation inputs.

[8] An LQG controller combines a Kalman filter and a Linear Quadratic Regulator (LQR), which minimizes a quadratic cost function.

[9] Advanced type of controller that predicts the evolution of the states of a system based on an existing model up until a certain prediction horizon H_p. The estimated state evolution is then used to solve an optimization problem, and the best control action found by minimization of a cost function J is applied to the system.

3.3 Dynamic Mode Decomposition Architecture in Fluid Dynamics

In the DMD framework, data is collected from a dynamical system:

$$\frac{d\boldsymbol{x}}{dt} = \mathbf{f}(\boldsymbol{x}; t; \mu) \tag{3.6}$$

where $\boldsymbol{x}(t) \in \mathbb{R}^n$ is a vector representing the state[10] of the dynamical system at time t, μ contains parameters of the system, and $\mathbf{f}(\cdot)$ represents the dynamics. Resorting to the equation free perspective, data measurements of the system alone are used to approximate the dynamics. The approximated linear dynamical system can be constructed in discrete time, sampling the analogous continuous time system every time interval Δt:

$$\boldsymbol{x}_{k+1} = \mathbf{A}\boldsymbol{x}_k \tag{3.7}$$

The relation in (3.7) puts in evidence the inherent assumption that a linear operator $\mathbf{A} \in \mathbb{R}^{n\times n}$ is responsible for the evolution of the system at each time instant. This operator $\mathbf{A}$ is found such that it optimally fits the measured trajectory $\boldsymbol{x}_k$ for $k = 1, 2, \ldots, m$. The word trajectory refers to the evolution of the state of the dynamical system over time. For this purpose, the *snapshot methodology* is used, where the information of interest from the flow at a certain instant is saved.

When analysing a time series of data containing the velocity field at a grid of spatial locations—as is the case for a wind farm—it is beneficial to take the three dimensional vector field data at time k and flatten into a single column vector. Consider a three dimensional flow field of a wake, where the wind velocity components u, v and w are sampled at every point of a three dimensional grid at a specific instant in time, as represented in Fig. 3.1. For the three dimensional vector field, $\boldsymbol{x}$ denotes the flow variable of interest, $\boldsymbol{r}$ the spatial coordinate, the index k and the kth time step. For a generalized grid with q, p and o points in the three cartesian directions, a snapshot of the flow field at one specific time instant is represented in Fig. 3.1.

The flattened resulting column vector $\boldsymbol{x}_k \in \mathbb{R}^{p\cdot q\cdot o}$, following the example in Fig. 3.1 containing the exact same information, is referred to as a *snapshot* of data.

[10] The states of a system are the variables which determine the evolution of the system. This concept, as well as the type of model used to represent the dynamical system, is thoroughly discussed in Chap. 5.

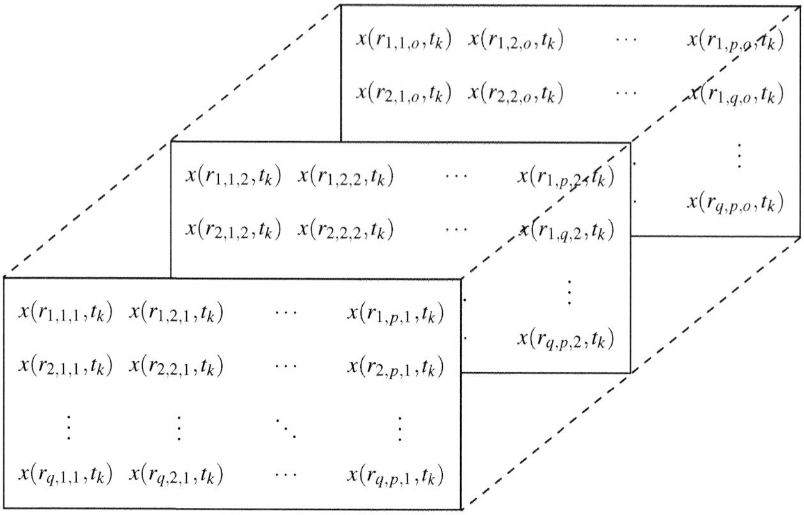

Fig. 3.1 A snapshot of the three dimensional field of a fluid in a three dimensional cartesian grid in a unique time instant

$$\boldsymbol{x}_k = \begin{bmatrix} x(r_{1,1,1}, t_k) \\ x(r_{1,2,1}, t_k) \\ \vdots \\ x(r_{2,1,1}, t_k) \\ x(r_{q,p,1}, t_k) \\ \vdots \\ x(r_{q,p,o}, t_k) \end{bmatrix} \tag{3.8}$$

The snapshots at each time instant are then combined to form the *data matrix* $\mathbf{X}_{full}$:

$$\mathbf{X}_{full} = \begin{bmatrix} | & | & & | \\ \boldsymbol{x}_1 & \boldsymbol{x}_2 & \cdots & \boldsymbol{x}_m \\ | & | & & | \end{bmatrix} \tag{3.9}$$

The number of rows n of the data matrix is equal to the number of spatial points saved per time snapshot. The number of columns m of the data matrix is equal to the number of snapshots taken, which in turn corresponds to the number of instants the flow field is sampled. The data matrix is then divided into two matrices in order to approximate the operator **A** in Eq. (3.7), where **X'** is the time-shifted snapshot matrix of **X**, as Eq. (3.10) clarifies.

$$\mathbf{X} = \begin{bmatrix} | & | & & | \\ \boldsymbol{x}_1 & \boldsymbol{x}_2 & \cdots & \boldsymbol{x}_{m-1} \\ | & | & & | \end{bmatrix} \qquad \mathbf{X}' = \begin{bmatrix} | & | & & | \\ \boldsymbol{x}_2 & \boldsymbol{x}_3 & \cdots & \boldsymbol{x}_m \\ | & | & & | \end{bmatrix} \tag{3.10}$$

> The snapshot methodology is at the core of DMD. The information of the system—all data gathered from the system when sampled—at a certain time instant k is gathered and stored in a single column vector, a snapshot. All snapshots are then combined into a single matrix, which contains the complete information resulting from the sampling of the system.

Equation (3.7) can now be rewritten in terms of the data snapshots:

$$\mathbf{X}' = \mathbf{A}\mathbf{X} \tag{3.11}$$

The best fit $\mathbf{A}$ matrix is given by

$$\mathbf{A} = \mathbf{X}'\mathbf{X}^{\dagger} \tag{3.12}$$

where † is the Moore-Penrose pseudoinverse. This solution minimizes the error $\|\mathbf{X}' - \mathbf{A}\mathbf{X}\|_F$, where $\|\cdot\|_F$ is the Frobenius norm [32], described in Annex A.1. Hence, DMD can be thought of as a least square fit,[11] or regression, of a best fit linear dynamical system $\boldsymbol{x}_{k+1} = \mathbf{A}\boldsymbol{x}_k$ for the data sampled. Matrix $\mathbf{A}$ is constructed so that $\|\boldsymbol{x}_{k+1} - \mathbf{A}\boldsymbol{x}_k\|_2$ is minimized across all snapshots, where $\|\cdot\|_2$ is the l_2 norm of a vector, described in Annex A.1.

In practice, due to the large state dimension n, matrix $\mathbf{A}$ becomes intractable to analyse directly, so DMD attempts to fit the snapshots in time using a low-rank matrix which captures the dynamics of the dataset. A low order representation of the states $\boldsymbol{x}_k$ can be written as

$$\tilde{\boldsymbol{x}}_k = \mathbf{Q}^*\boldsymbol{x}_k \tag{3.13}$$

where $\mathbf{Q}$ is a generic orthonormal basis that forms the projection subspace and * designates the conjugate transpose.

In order to find the best fit linear operator $\mathbf{A}$ and its corresponding low-rank matrix, the standard DMD algorithm described in [21, 25, 33] proceeds as follows:

[11] A least squares regression is a regression technique where a best fit line or vector is found by minimizing the sum of squares of the error between the model and the data.

1. *The Singular Value Decomposition (SVD) of X is taken*:

$$\mathbf{X} = \mathbf{U}\boldsymbol{\Sigma}\mathbf{V}^* = \begin{bmatrix}\tilde{\mathbf{U}} \; \tilde{\mathbf{U}}_{\text{rem}}\end{bmatrix} \begin{bmatrix}\tilde{\boldsymbol{\Sigma}} & 0 \\ 0 & \tilde{\boldsymbol{\Sigma}}_{\text{rem}}\end{bmatrix} \begin{bmatrix}\tilde{\mathbf{V}}^* \\ \tilde{\mathbf{V}}^*_{\text{rem}}\end{bmatrix} \tag{3.14}$$

where: $\mathbf{U} \in \mathbb{R}^{n \times n}$, $\boldsymbol{\Sigma} \in \mathbb{R}^{n \times m-1}$, $\mathbf{V}^* \in \mathbb{R}^{m-1 \times m-1}$ and $\tilde{\mathbf{U}} \in \mathbb{R}^{n \times r}$, $\tilde{\boldsymbol{\Sigma}} \in \mathbb{R}^{r \times r}$, $\tilde{\mathbf{V}}^* \in \mathbb{R}^{r \times m-1}$. The subscript $_{\text{rem}}$ indicates the remaining $m - 1 - r$ singular values. By appropriately choosing a truncation value r of the number of singular values to retain, it is possible to eliminate the remaining $_{\text{rem}}$ terms and find a reduced SVD approximation of $\mathbf{X}$.

$$\mathbf{X} \approx \tilde{\mathbf{U}}\tilde{\boldsymbol{\Sigma}}\tilde{\mathbf{V}}^* \tag{3.15}$$

For simplicity purposes, the truncated SVD approximation of $\mathbf{X}$ will be referred simply as

$$\mathbf{X} \approx \mathbf{U}\boldsymbol{\Sigma}\mathbf{V}^* \tag{3.16}$$

2. *Matrix A in (3.11) is computed by substituting the SVD of X (3.16) in (3.12)*:
The pseudoinverse can now be accomplished since $\boldsymbol{\Sigma}$ is squared. By using the known result of the pseudoinverse of the SVD:

$$\mathbf{A} \approx \mathbf{X}'(\mathbf{U}\boldsymbol{\Sigma}\mathbf{V}^*)^\dagger \approx \mathbf{X}'\mathbf{V}\boldsymbol{\Sigma}^{-1}\mathbf{U}^* \tag{3.17}$$

3. *Matrix A from (3.17) is projected onto a subspace and a low dimension linear model $\tilde{\mathbf{A}}$ is obtained*:
Matrix $\mathbf{A}$ in (3.17) is high dimensional of size $n \times n$, and an eigenvalue analysis to undercover the underlying dynamics may prove to be unfeasible due to the associated high computational cost. Consequently, a more compact and computationally efficient model can be found, assuming that $r \ll n$, by projecting the full state onto a linear subspace of dimension r. A convenient transformation has already been computed via the SVD of $\mathbf{X}$, and the projection subspace $\mathbf{Q}$ corresponds to the left singular vectors $\mathbf{U}$, the POD modes.

$$\mathbf{A} \approx \tilde{\mathbf{A}} = \mathbf{U}^*\mathbf{A}\mathbf{U} = \mathbf{U}^*\mathbf{X}'\mathbf{V}\boldsymbol{\Sigma}^{-1}\mathbf{U}^*\mathbf{U} \tag{3.18}$$

The columns of $\mathbf{U}$ are orthonormal due to the properties of the SVD, which means that $\mathbf{U}^*\mathbf{U} = \mathbf{I}$. Taking this into account, (3.18) reduces to

$$\tilde{\mathbf{A}} = \mathbf{U}^*\mathbf{X}'\mathbf{V}\boldsymbol{\Sigma}^{-1} \tag{3.19}$$

Matrix $\tilde{\mathbf{A}}$ defines the low dimensional linear model of the dynamical system on POD coordinates:

$$\tilde{\boldsymbol{x}}_{k+1} = \tilde{\mathbf{A}}\tilde{\boldsymbol{x}}_k \tag{3.20}$$

The high dimensional states $\boldsymbol{x}_k$ can be reconstructed from the low dimension states $\tilde{\boldsymbol{x}}_k$ by $\boldsymbol{x}_k = \mathbf{U}\tilde{\boldsymbol{x}}_k$.

4. *The eigendecomposition of* $\tilde{\mathbf{A}}$ *is taken:*

$$\tilde{\mathbf{A}}\mathbf{W} = \mathbf{W}\boldsymbol{\Lambda} \tag{3.21}$$

where columns of $\mathbf{W}$ are eigenvectors and $\boldsymbol{\Lambda}$ is a diagonal matrix containing the corresponding eigenvalues λ_k. These can be investigated for fundamental properties of the underlying system such as growth modes and resonance frequencies.

5. *The eigendecomposition of A is reconstructed:*
To approximate the eigenvalues and eigenvectors of $\mathbf{A}$ without its explicit computation, an approximation is calculated based on the so-called *DMD modes*. The eigenvalues of $\mathbf{A}$ are given by $\boldsymbol{\Lambda}$, and the eigenvectors are approximated by the dynamic modes, which correspond to the columns of $\boldsymbol{\Phi}$ in (3.22), as originally proposed in [23] and figuring in the standard DMD algorithm in [25]:

$$\boldsymbol{\Phi} = \mathbf{U}\mathbf{W} \tag{3.22}$$

These modes in (3.22) are referred to as *projected DMD modes*. Another formulation is used to compute the DMD modes:

$$\boldsymbol{\Phi} = \mathbf{X}'\mathbf{V}\boldsymbol{\Sigma}^{-1}\mathbf{W} \tag{3.23}$$

The modes in (3.23) are often called *exact DMD modes*, as it was proven in [25] that these are the exact eigenvectors of the matrix $\mathbf{A}$. The projected DMD modes will tend to converge if $\mathbf{X}$ and $\mathbf{X'}$ have the same column space [21].

In the context of fluid mechanics, the modes are spatial fields that often identify coherent structures in the flow. The correspondent eigenvalues define growth or decay rates and oscillation frequencies for each mode. Taken together, the DMD modes and eigenvalues describe the dynamics observed in the time series in terms of oscillatory components. This technique provides a powerful tool to have a better understanding of the different phenomena that are present in the wake of a wind turbine.

3.4 Dynamic Mode Decomposition with Control

One of the algorithms following the original DMD is Dynamic Mode Decomposition with control (DMDc), as fully described in [33]. The latter modifies the basic assumption of DMD, in the sense that the linear dynamical system connecting the future state $\boldsymbol{x}_{k+1}$ now relies not only on the current state $\boldsymbol{x}_k$, as expressed in (3.7) but also on the current control $\boldsymbol{u}_k$, given by the following relation:

$$\boldsymbol{x}_{k+1} = \mathbf{A}\boldsymbol{x}_k + \mathbf{B}\boldsymbol{u}_k \tag{3.24}$$

where $\boldsymbol{x} \in \mathbb{R}^n$, $\boldsymbol{u} \in \mathbb{R}^q$, $\mathbf{A} \in \mathbb{R}^{n\times n}$ and $\mathbf{B} \in \mathbb{R}^{n\times q}$. As explained in the original article [33], DMDc allows to understand the characteristics of complex systems by helping discover the underlying dynamics without the confounding effect of external control. In addition, the method also quantifies the effect of control inputs on the state of the system.

The DMDc algorithm operates in a similar fashion as the standard DMD. Besides collecting the snapshot matrices as in (3.10), the sequence of control input snapshots is assembled:

$$\boldsymbol{\Upsilon} = \begin{bmatrix} | & | & & | \\ \boldsymbol{u}_1 & \boldsymbol{u}_2 & \cdots & \boldsymbol{u}_{m-1} \\ | & | & & | \end{bmatrix} \tag{3.25}$$

The relation (3.24) can be rewritten in terms of the snapshot matrices:

$$\mathbf{X}' \approx \mathbf{A}\mathbf{X} + \mathbf{B}\boldsymbol{\Upsilon} \tag{3.26}$$

Assuming both operators **A** and **B** to be unknown, i.e. there exists no knowledge about how the control affects the system, (3.26) is manipulated giving the following representation:

$$\mathbf{X}' \approx \begin{bmatrix} \mathbf{A} \ \mathbf{B} \end{bmatrix} \begin{bmatrix} \mathbf{X} \\ \boldsymbol{\Upsilon} \end{bmatrix} = \mathbf{G}\boldsymbol{\Omega} \tag{3.27}$$

where $\boldsymbol{\Omega}$ contains both the state and control snapshot information. Similarly to the standard DMD, let the DMDc be presented in an algorithmic fashion, where the goal is to seek the low order best fit solution of operator **G**, as described in [33]:

1. *The SVD of the input space* $\boldsymbol{\Omega}$ *is computed:*
   The SVD of $\boldsymbol{\Omega}$ is taken with a certain truncation value p:=

$$\boldsymbol{\Omega} \approx \tilde{\mathbf{U}}\tilde{\boldsymbol{\Sigma}}\tilde{\mathbf{V}}^* \tag{3.28}$$

An approximation for operator **G** is now found by taking the pseudoinverse of the SVD of $\boldsymbol{\Omega}$: $\mathbf{G} \approx \mathbf{X}'\tilde{\mathbf{V}}\tilde{\boldsymbol{\Sigma}}^{-1}\tilde{\mathbf{U}}^*$, where $\mathbf{G} \in \mathbb{R}^{n\times(n+q)}$. By breaking the linear operator $\tilde{\mathbf{U}}$ in two separate components, an explicit approximation of the operators **A** and **B** is found:

$$[\mathbf{A}, \mathbf{B}] \approx [\mathbf{X}'\tilde{\mathbf{V}}\tilde{\boldsymbol{\Sigma}}^{-1}\tilde{\mathbf{U}}_1^*, \mathbf{X}'\tilde{\mathbf{V}}\tilde{\boldsymbol{\Sigma}}^{-1}\tilde{\mathbf{U}}_2^*] \tag{3.29}$$

where $\tilde{\mathbf{U}}_1 \in \mathbb{R}^{n\times p}$, $\tilde{\mathbf{U}}_2 \in \mathbb{R}^{q\times p}$, $\tilde{\mathbf{U}} = \begin{bmatrix} \tilde{\mathbf{U}}_1^* \ \tilde{\mathbf{U}}_2^* \end{bmatrix}^T$, and p is the truncation value of the SVD in (3.28).

2. *The SVD of the output space X' is computed:* The SVD of **X'** is needed to find a projection subspace matrix **Q** to compute the low order model. DMD makes use of the truncated left singular values **U** of **X** for such purpose. However, DMDc cannot make use of the left singular vectors $\tilde{\mathbf{U}}$ of $\boldsymbol{\Omega}$ as it is defined for the input space, which now includes both the state measurements and the exogenous inputs, as mentioned in [21, 33]. Consequently, the data matrix of the output space **X'** is used: $\mathbf{X}' \approx \hat{\mathbf{U}}\hat{\boldsymbol{\Sigma}}\hat{\mathbf{V}}^*$, where r is the truncation value, $\hat{\mathbf{U}} \in \mathbb{R}^{n \times r}$, $\hat{\boldsymbol{\Sigma}} \in \mathbb{R}^{r \times r}$ and $\hat{\mathbf{V}}^* \in \mathbb{R}^{r \times m-1}$. For a tractable low order model to be found, usually $r < p$.

3. *An approximation of the low order operator G is computed:*
The reduced order approximation of operators **A** and **B** is then found to be:

$$\begin{aligned} \tilde{\mathbf{A}} &= \hat{\mathbf{U}}^*\mathbf{A}\hat{\mathbf{U}} = \hat{\mathbf{U}}^*\mathbf{X}'\tilde{\mathbf{V}}\tilde{\boldsymbol{\Sigma}}^{-1}\tilde{\mathbf{U}}_1^*\hat{\mathbf{U}} \\ \tilde{\mathbf{B}} &= \hat{\mathbf{U}}^*\mathbf{B} = \hat{\mathbf{U}}^*\mathbf{X}'\tilde{\mathbf{V}}\tilde{\boldsymbol{\Sigma}}^{-1}\tilde{\mathbf{U}}_2^* \end{aligned} \tag{3.30}$$

where $\tilde{\mathbf{A}} \in \mathbb{R}^{r \times r}$ and $\tilde{\mathbf{B}} \in \mathbb{R}^{r \times q}$. The reduced order model can now be written as

$$\tilde{\boldsymbol{x}}_{k+1} = \tilde{\mathbf{A}}\tilde{\boldsymbol{x}}_k + \tilde{\mathbf{B}}\boldsymbol{u}_k \tag{3.31}$$

4. *The eigenvalue decomposition of $\tilde{\mathbf{A}}$ is performed:*

$$\tilde{\mathbf{A}}\mathbf{W} = \mathbf{W}\boldsymbol{\Lambda} \tag{3.32}$$

where columns of **W** are eigenvectors and $\boldsymbol{\Lambda}$ is a diagonal matrix containing the corresponding eigenvalues λ_k.

5. *The dynamic modes of operator A are reconstructed:*
The transformation from eigenvectors to dynamic modes of **A** is slightly modified and is given by

$$\boldsymbol{\Phi} = \mathbf{X}'\tilde{\mathbf{V}}\tilde{\boldsymbol{\Sigma}}^{-1}\tilde{\mathbf{U}}_1^*\hat{\mathbf{U}}\mathbf{W} \tag{3.33}$$

The theoretical mathematical framework of DMDc proves its value in terms of producing accurate reduced order models from complex systems with exogenous forcing. Nevertheless, it is developed for a specific regime, where the number of state measurements is much larger than the rank of the underlying system [21]. For many engineering applications, the number of measurements is typically smaller than the number of states to describe the input-output models. Such challenge is addressed in the next variation of the DMD algorithm.

Despite the mentioned feature incapacitating DMDc of being exploited in certain engineering applications, it lays the grounds for the development of better suited algorithms. Furthermore, its applicability is wide. In the context of epidemiology, for example, the extension of DMD to DMDc helps planning eradication campaigns by identifying how such inputs drive the disease propagation [26].

3.5 Input-Output Dynamic Mode Decomposition

In this section, an extension of the DMD approach to include inputs and outputs is introduced, based on the work in [29, 34], where reduced order models are built to model the wind farm flow based on a medium fidelity and a high fidelity simulation, respectively.

In Sect. 3.4, DMD is studied in the context of control. This approach projects the full-order model onto the output subspace, where the inputs are accounted for when fitting the data to a linear system. Input-Output Dynamic Mode Decomposition (IODMD), in its turn, combines DMD with standard subspace identification, often used in the controls literature [35], to obtain an input-output characterization of the system.

Input-Output Dynamic Mode Decomposition (IODMD) is a useful extension of the Dynamic Mode Decomposition algorithm. By combining it with subspace identification, both inputs and outputs can be taken into consideration when computing the ROM.

In addition to the relation in (3.24), a second is added to account for the mapping between the state measurement and the outputs:

$$\begin{aligned} \boldsymbol{x}_{k+1} &= \mathbf{A}\boldsymbol{x}_k + \mathbf{B}\boldsymbol{u}_k \\ \boldsymbol{y}_k &= \mathbf{C}\boldsymbol{x}_k + \mathbf{D}\boldsymbol{u}_k \end{aligned} \tag{3.34}$$

where $\boldsymbol{x}_k \in \mathbb{R}^n$, $\boldsymbol{u}_k \in \mathbb{R}^q$ and $\boldsymbol{y}_k \in \mathbb{R}^l$ are the state, input and output vectors, respectively. Matrices $\mathbf{A} \in \mathbb{R}^{n\times n}$, $\mathbf{B} \in \mathbb{R}^{n\times q}$, $\mathbf{C} \in \mathbb{R}^{l\times n}$ and $\mathbf{D} \in \mathbb{R}^{l\times q}$ are the state matrices that together form the *state space representation*, a common mathematical model of a physical system used in control engineering, further explored in Sect. 4.1.2.

The sequence of output snapshots is now assembled, as well as the input and states snapshots—also previously gathered in the DMDc and DMD methodologies.

$$\mathbf{Y} = \begin{bmatrix} | & | & & | \\ \boldsymbol{y}_1 & \boldsymbol{y}_2 & \cdots & \boldsymbol{y}_{m-1} \\ | & | & & | \end{bmatrix} \tag{3.35}$$

Rewriting (3.34) in terms of the snapshot matrices:=

$$\begin{aligned} \mathbf{X}' &\approx \mathbf{AX} + \mathbf{B}\boldsymbol{\Upsilon} \\ \mathbf{Y} &\approx \mathbf{CX} + \mathbf{D}\boldsymbol{\Upsilon} \end{aligned} \tag{3.36}$$

The relation in (3.36) can again be rewritten in a more compact format, which allows to find the low order best fit operators of **A**, **B**, **C**, **D**:

$$\begin{bmatrix} \mathbf{X}' \\ \mathbf{Y} \end{bmatrix} = \begin{bmatrix} \mathbf{A} & \mathbf{B} \\ \mathbf{C} & \mathbf{D} \end{bmatrix} \begin{bmatrix} \mathbf{X} \\ \boldsymbol{\Upsilon} \end{bmatrix} \tag{3.37}$$

The state is then projected onto the subspace defined by an orthonormal basis specified by the columns of matrix **Q**, following the same approach as in (3.13). The low order state matrices are then $\tilde{\mathbf{A}} = \mathbf{Q}^*\mathbf{AQ}$, $\tilde{\mathbf{B}} = \mathbf{Q}^*\mathbf{B}$, $\tilde{\mathbf{C}} = \mathbf{CQ}$ and $\tilde{\mathbf{D}} = \mathbf{D}$. These equivalences can be rewritten in terms of the high order matrices **A**, **B**, **C** and **D**: $\mathbf{A} = \mathbf{Q}\tilde{\mathbf{A}}\mathbf{Q}^*$, $\mathbf{B} = \mathbf{Q}\tilde{\mathbf{B}}$, $\mathbf{C} = \tilde{\mathbf{C}}\mathbf{Q}^*$ and $\mathbf{D} = \tilde{\mathbf{D}}$. Substituting in (3.37) and rearranging:

$$\begin{bmatrix} \mathbf{X}' \\ \mathbf{Y} \end{bmatrix} = \begin{bmatrix} \mathbf{Q} & \mathbf{0} \\ \mathbf{0} & \mathbf{I} \end{bmatrix} \begin{bmatrix} \tilde{\mathbf{A}} & \tilde{\mathbf{B}} \\ \tilde{\mathbf{C}} & \tilde{\mathbf{D}} \end{bmatrix} \begin{bmatrix} \mathbf{Q}^* & \mathbf{0} \\ \mathbf{0} & \mathbf{I} \end{bmatrix} \begin{bmatrix} \mathbf{X} \\ \boldsymbol{\Upsilon} \end{bmatrix} \tag{3.38}$$

A suboptimal but useful choice for the projection space is given by the POD modes of **X** [34], as in standard DMD. By taking the SVD of **X** given by $\mathbf{X} \approx \mathbf{U}\boldsymbol{\Sigma}\mathbf{V}^*$ with a truncation value r, **U** is chosen as **Q**.

The reduced order state matrices are obtained by minimizing the error of the Frobenius norm:

$$\min_{\begin{bmatrix} \tilde{\mathbf{A}} & \tilde{\mathbf{B}} \\ \tilde{\mathbf{C}} & \tilde{\mathbf{D}} \end{bmatrix}} \left\| \begin{bmatrix} \mathbf{X}' \\ \mathbf{Y} \end{bmatrix} - \begin{bmatrix} \mathbf{U} & \mathbf{0} \\ \mathbf{0} & \mathbf{I} \end{bmatrix} \begin{bmatrix} \tilde{\mathbf{A}} & \tilde{\mathbf{B}} \\ \tilde{\mathbf{C}} & \tilde{\mathbf{D}} \end{bmatrix} \begin{bmatrix} \mathbf{U}^* & \mathbf{0} \\ \mathbf{0} & \mathbf{I} \end{bmatrix} \begin{bmatrix} \mathbf{X} \\ \boldsymbol{\Upsilon} \end{bmatrix} \right\|_F^2 \tag{3.39}$$

where $\tilde{\mathbf{A}} \in \mathbb{R}^{r \times r}$, $\tilde{\mathbf{B}} \in \mathbb{R}^{r \times q}$, $\tilde{\mathbf{C}} \in \mathbb{R}^{l \times r}$ and $\tilde{\mathbf{D}} \in \mathbb{R}^{l \times q}$. The reduced order state matrices are then obtained by least squares, taking into account that the reduced order representations of **X** and **X'** are respectively $\mathbf{U}^*\mathbf{X}'$ and $\boldsymbol{\Sigma}\mathbf{V}$:

$$\Theta_{\text{IODMD}} = \begin{bmatrix} \tilde{\mathbf{A}} & \tilde{\mathbf{B}} \\ \tilde{\mathbf{C}} & \tilde{\mathbf{D}} \end{bmatrix} = \begin{bmatrix} \mathbf{U}^*\mathbf{X}' \\ \mathbf{Y} \end{bmatrix} \begin{bmatrix} \boldsymbol{\Sigma}\mathbf{V} \\ \boldsymbol{\Upsilon} \end{bmatrix}^{\dagger} \tag{3.40}$$

Using the definition of the pseudoinverse for a general matrix **X** and under the assumption that **X** has linearly independent rows, $\mathbf{X}^{\dagger}$ can be written as $\mathbf{X}^* \, (\mathbf{X}^* \, \mathbf{X})^{-1}$. This result allows to rewrite equation (3.40) in a more explicit format, which will be of use later in this chapter in Sect. 3.7:

$$\Theta_{\text{IODMD}} = \begin{bmatrix} \tilde{\mathbf{A}} & \tilde{\mathbf{B}} \\ \tilde{\mathbf{C}} & \tilde{\mathbf{D}} \end{bmatrix} = \begin{bmatrix} \mathbf{U}^*\mathbf{X}' \\ \mathbf{Y} \end{bmatrix} \begin{bmatrix} \boldsymbol{\Sigma}\mathbf{V} \\ \boldsymbol{\Upsilon} \end{bmatrix}^T \begin{bmatrix} \boldsymbol{\Sigma}\mathbf{V}\boldsymbol{\Sigma}^T\mathbf{V}^T & \boldsymbol{\Sigma}\mathbf{V}\boldsymbol{\Upsilon}^T \\ \boldsymbol{\Upsilon}\boldsymbol{\Sigma}^T\mathbf{V}^T & \boldsymbol{\Upsilon}\boldsymbol{\Upsilon}^T \end{bmatrix}^{-1} \tag{3.41}$$

The reduced order model can now be written as

$$\begin{aligned} \tilde{\boldsymbol{x}}_{k+1} &= \tilde{\mathbf{A}}\tilde{\boldsymbol{x}}_k + \tilde{\mathbf{B}}\boldsymbol{u}_k \\ \boldsymbol{y}_k &= \tilde{\mathbf{C}}\tilde{\boldsymbol{x}}_k + \tilde{\mathbf{D}}\boldsymbol{u}_k \end{aligned} \tag{3.42}$$

It is important to emphasize a point stated in [29]: The ROM arrived at (3.42) is not a *black box model*[12] as the state of the ROM can be used to approximate the full-order state, by $\boldsymbol{x}_k = \mathbf{U}\tilde{\boldsymbol{x}}_k$.

The code snippet Listing 3.1 is a MATLAB implementation of the IODMD algorithm and illustrates how it is possible to compute a ROM with 100 singular values (merely as an example) using post-processed data and inbuilt MATLAB functions.

Listing 3.1 Implementation of Input-Output Dynamic Mode Decomposition in MATLAB

```
function [ROM]=iodmd(states, Inputs, Outputs)

    %% INPUT ARGUMENTS
    %states:  states to build DMD matrices
    %inputs:  [U1 U2 U3 ... Un];
    %outputs: [Y1 Y2 Y3 .... Yn];

    %define necessary matrices for IODMD
    X       = states(:,1:end-1);        %snapshot matrix
    X_p     = states(:,2:end);          %time-shifted snapshot matrix

    out     = Outputs(:,1:end-1);       %output data matrix
    out_p   = Outputs(:,2:end);         %ime shifted output data matrix

    inp     =Inputs(:,1:end-1);         %input data matrix
    inp_p   =Inputs(:,2:end);           %time shifted input data matrix

    %define truncation value for the SVD
    r=100

    %Perform SVD of X, returning r largest singular values.
    %Extract the left singular vectors (U),
    %singular values (S) and right singular vectors (V).
    [U,S,V]=svds(X,r);

    %Use IODMD solution
    matr=[ U'*X_p;out]*pinv([S*V';inp]);

    %Fetch A,B,C,D matrices
    A=matr(1:r,1:r);
    B=matr(1:r,r+1:end);
    C=matr(r+1:end, 1:r);
    D=matr(r+1:end, r+1:end);

    %compute state space Reduced Order Model
    ROM=ss(A,B,C,D,2);

end
```

[12] System which can be solemnly viewed in terms of its inputs and outputs, without any knowledge of its internal working.

3.6 Extended Dynamic Mode Decomposition

It has been shown that a non-linear system can be described by superposition of modes whose dynamics are governed by eigenvalues. However, at a first glance, it may seem dubious that such description of a complex non-linear system provides accurate information and that DMD is indeed applicable to non-linear systems. Nonetheless, evidence exists supporting DMD applicability due to its connections to what is termed the *Koopman operator*, as shown in [36]. In this subsection, as a result of its relevance, this operator and its connections to DMD are explored along with variations it incites to the already seen frameworks.

3.6.1 Koopman Operator Theory

The Koopman operator was introduced in [37]. It has recently gathered attention given the amount of data available, the existence of complex non-linear system of interest to model and its relevance in real work applications.

Succinctly, it is defined as an infinite-dimensional linear operator that evolves fields of physical observables—scalar functions—defined on a state space of a dynamical system.

More explicitly, and following the nomenclature used in [38], let the generic dynamical system in Eq. (3.6) be again considered, and defined on a state space $\mathcal{M}$. This induces a discrete time dynamical system given by the flow map $\mathcal{F}_t : \mathcal{M} \rightarrow \mathcal{M}$, represented in Fig. 3.2.

$$\boldsymbol{x}_{k+1} = \mathcal{F}_t(\boldsymbol{x}_k) \tag{3.43}$$

A function defined as $\boldsymbol{g} : \mathcal{M} \rightarrow \mathbb{R}$ is called an *observable* of the system, as represented in Fig. 3.2. The Koopman operator $\mathcal{K}_t$ acts on observable functions g as

$$\mathcal{K}_t = g \circ \mathcal{F}_t \tag{3.44}$$

where $\circ$ is the composition operator, so that:

$$\mathcal{K}_t \boldsymbol{g}(\boldsymbol{x}_k) = \boldsymbol{g}(\mathcal{F}_t(\boldsymbol{x}_k)) = \boldsymbol{g}(\boldsymbol{x}_{k+1}) \tag{3.45}$$

$$\boldsymbol{g}(\boldsymbol{x}_{k+1}) = \mathcal{K}_t \boldsymbol{g}(\boldsymbol{x}_k) \tag{3.46}$$

It is now evident in (3.46) that the observation of the state, or, in other words, a function of the measured state variable, is advanced to the next time step by the Koopman operator. This is true for *any* observable function and for *any* point $\boldsymbol{x}_k \in$ **M**, hence the infinite dimension. The linearity property follows from its definition in (3.44).

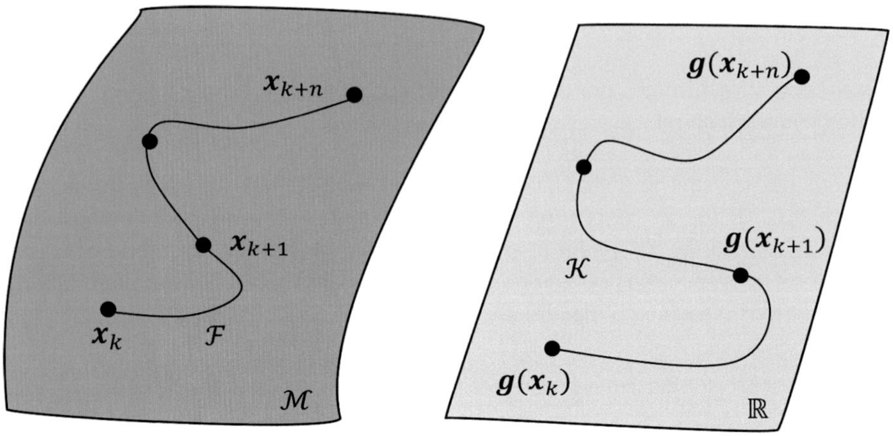

Fig. 3.2 Schematic representation of the Koopman operator. On the left, the (non-linear) state evolution of a certain dynamical system defined on a state space $\mathscr{M}$, dictated by mapping $\mathscr{F}$. On the right, the evolution of the observables of the system in a linear fashion by the Koopman operator $\mathscr{K}$. Image adapted from [38] with the authors permission

The spectral decomposition[13] of the Koopman operator is instrumental in representing solutions to a dynamical system of interest. Considering the eigenvalue problem:

$$\mathscr{K}\varphi_k = \lambda_k \varphi_k \tag{3.47}$$

The functions $\varphi_k(\boldsymbol{x})$ are called Koopman eigenfunctions, and they define a set of intrinsic measurement coordinates, on which it is possible to advance these measurements with a linear dynamical system [38]. The new measurement space formed by a vector of observables $\boldsymbol{g}(\boldsymbol{x})$ can be expressed in terms of the Koopman eigenfunctions:

$$\boldsymbol{g}(\boldsymbol{x}) = \begin{bmatrix} g_1(\boldsymbol{x}) \\ g_2(\boldsymbol{x}) \\ \vdots \\ g_p(\boldsymbol{x}) \end{bmatrix} = \sum_{k=1}^{\infty} \varphi_k(\boldsymbol{x}) \boldsymbol{v}_k \tag{3.48}$$

where $\boldsymbol{v}_k$ is the kth Koopman mode of the map $\mathscr{F}$ corresponding to the observable $\boldsymbol{g}$, associated with the kth Koopman eigenfunction, i.e. the weighting of each

[13] Spectral decomposition and eigendecomposition are two terms used almost interchangeably. Spectral theory simply refers to a broader set of theories which extend the eigenvector and eigenvalue theories to a broader theory concerning the structure of operator in various mathematical spaces.

observable on the eigenfunction. The expression in (3.48) can be interpreted as a linear combination of the eigenfunctions φ_k of $\mathscr{K}$ where $\boldsymbol{v}_k$ is the coefficient in the expansion [38]. The non-linear system can either be evolved in the original state space or in the measurement space as in (3.48).

$$\mathscr{K} g(\boldsymbol{x}) = \mathscr{K} \sum_{k=1}^{\infty} \varphi_k(\boldsymbol{x}) \boldsymbol{v}_k = \sum_{k=1}^{\infty} \mathscr{K} \varphi_k(\boldsymbol{x}) \boldsymbol{v}_k = \sum_{k=1}^{\infty} \lambda_k \varphi_k(\boldsymbol{x}) \boldsymbol{v}_k \tag{3.49}$$

Future solutions can be computed by multiplication with the Koopman eigenvalues, $\lambda_k \in \mathbb{C}$, which characterize the temporal behaviour of the corresponding Koopman mode $\boldsymbol{v}_k$.

The connections of DMD to the underlying non-linear dynamics of a system through Koopman operator theory were firstly established in [36], where it is shown that DMD produces approximations to eigenvalues of the Koopman operator, and their corresponding modes. A thorough mathematical definition of the Koopman operator and its properties as well as an overview of the rapidly growing work using both these concepts can be seen in [39].

Dynamic Mode Decomposition is thus a manifestation of Koopman theory for the specific case when the observable functions are the identity or a linear transformation of the underlying state space [21]:

$$g_{\mathrm{DMD}}(\boldsymbol{x}) = \boldsymbol{x} \tag{3.50}$$

Dynamic Mode Decomposition provides a tractable computational alternative, as the finite-dimensional matrices it yields are an approximation to the infinite-dimensional operator. In other words, DMD transforms the useful mathematical conception to a readily implementable computation.

The connections with the Koopman operator have given birth to variations of DMD that attempt to better approximate the Koopman eigenvalues and modes and hence improve the approximation of the full non-linear dynamics. The key idea is to apply DMD on the space of observables. For that purpose, the snapshot matrices in (3.10) are generalized and assembled with observables. Derivations of the classical DMD and IODMD motivated by the ideas of the Koopman operator are explored in the subsequent subsection.

3.6.2 Extended Input-Output Dynamic Mode Decomposition

The Extended Dynamic Mode Decomposition (EDMD) was introduced in [40], augmenting DMD to include a richer set of non-linear observables and as a consequence providing the ability of effectively analysing non-linear systems. For high dimensional systems where such extension might become too expensive computationally wise, techniques from machine learning can be used as shown in [41], thus benefiting from the richness of the non-linear observables while maintaining an acceptable computational cost.

The major challenge this approach faces concerns the choosing of the non-linear observables, as shown in [42, 43]. Furthermore, in the latter other drawbacks of EDMD are discussed, such as the loss of interpretability (common in machine learning schemes) and high variability of results (reiterating the importance of cross validation).

An alternative to the standard EDMD using machine learning techniques is to select a parsimonious set of observables based upon knowledge of the physics of the system. This means that knowledge of non-linear terms figuring in the Non-linear Partial Differential Equations (NLPDE) that govern the dynamics of complex systems is used to guide the selection of observables (analogous to the selection of *features* in the context of machine learning guided by expert knowledge).

An extensive overview of such approach is given in [43], where compelling evidence is presented: By judiciously selecting observables based on non-linear terms figuring in the Schrödinger NPDE, the deviations between DMD's approximation and true solution dramatically decrease (in about three orders of magnitude). Additionally, in [44], it is also shown that when combined with adequate choice of observables, the Koopman operator is able to map the non-linear underlying dynamics with the linear observable space, where the DMD can be implemented with good accuracy (and efficiency).

The linear representation provided by the Koopman operator and its computational implementable correspondent DMD are useful not only for prediction but also for control, as explored in the works of [45–47]. In the last two, a structured approach is given to derive linear models suited for predictive control based on the ideas of the Koopman operator. It boils down to a non-linear transformation of the data—termed the *lifting*—and a linear least squares problem in the lifted space.

In the works of [30, 48], these ideas are explored by using the Extended Input-Output Dynamic Mode Decomposition (EIODMD) approach, which combines the benefits of using non-linear observables with the subspace identification methods and least squares regression exploited in IODMD, allowing to derive tractable models for predictive control that embed the relevant dynamics. These results will be discussed in depth in the following chapters.

In [48], the EIODMD approach is tested, and it is seen that augmenting the vector of observations with other variables and non-linear combinations of variables is beneficial and leads to more accurate models. The models are not only better suited in terms of capturing the behaviour of the output (in this specific case, the generated power of a downstream turbine) but also in terms of the state space trajectory (in this case the behaviour of the upstream turbine wake). The increase in accuracy of the models depends naturally on the choosing of the variables used to augment the vector of observables. Note that this example used data from a high fidelity simulation where Axial Induction Control was tested in a two turbine set-up.

This is also an example of the first and second tasks of DMD mentioned earlier in Sect. 3.2.2. Furthermore, these previous results will be explored in great detail in the following sections of this book.

In [30] EIODMD is also tested. However in this case the application goes one step further from modelling, and the system studied is a two turbine set where Wake Redirection Control is tested. The model found is then used to design a Model Predictive Controller (MPC), an advanced control strategy where the future state of the system is forecasted up until a certain prediction horizon H_p. This forecast is then used to solve an optimization problem where the optimal set of control actions is computed. In the former article, the goal was to demand a pre-specified combined power production from the two turbines. The optimization problem takes this objective into account when solving for the best set of control actions for the specified H_p, and the first input of these is applied to the system. In the context of the work in [30], the control action is the yaw misalignment setpoint of the upstream turbine. It is seen that the designed MPC allows the combined power production to be tracked and leads to a collaborative behaviour of both turbines by successfully taking into account the delayed wake interaction between the two.

The Koopman Operator motivates a set of modifications to the already seen algorithms. The underlying hypothesis is that there may exist a better set of coordinates to evolve the system's dynamics than the ones initially thought.

In practical terms, the formulation of the problem is similar to IODMD, where the goal is to find the ROM that best fits the gathered data. The snapshot matrices are then lifted (with a slight different notation to match the current literature, albeit maintaining its original meaning):

$$\mathbf{X} = \begin{bmatrix} | & | & & | \\ \boldsymbol{g}(\boldsymbol{x}_1) & \boldsymbol{g}(\boldsymbol{x}_2) & \cdots & \boldsymbol{g}(\boldsymbol{x}_{m-1}) \\ | & | & & | \end{bmatrix} \qquad \mathbf{X}' = \begin{bmatrix} | & | & & | \\ \boldsymbol{g}(\boldsymbol{x}_2) & \boldsymbol{g}(\boldsymbol{x}_3) & \cdots & \boldsymbol{g}(\boldsymbol{x}_m) \\ | & | & & | \end{bmatrix} \tag{3.51}$$

where $\boldsymbol{g}(\boldsymbol{x}_k) = \begin{bmatrix} g_1(x_k) & g_2(x_k) & \cdots & g_p(x_k) \end{bmatrix}^T$. The vector $\boldsymbol{g}$ is the vector of observables, where each function g_p is a transformation of a certain variable x which has a certain value at time instant k.

The algorithm proceeds in a similar fashion as before. The projection subspace $\mathbf{Q}$ corresponds to the left singular vectors resulting from the SVD of $\mathbf{X}$:

$$\min_{\begin{bmatrix} \tilde{\mathbf{A}} & \tilde{\mathbf{B}} \\ \tilde{\mathbf{C}} & \tilde{\mathbf{D}} \end{bmatrix}} \left\| \begin{bmatrix} \mathbf{X}' \\ \mathbf{Y} \end{bmatrix} - \begin{bmatrix} \mathbf{U} & \mathbf{0} \\ \mathbf{0} & \mathbf{I} \end{bmatrix} \begin{bmatrix} \tilde{\mathbf{A}} & \tilde{\mathbf{B}} \\ \tilde{\mathbf{C}} & \tilde{\mathbf{D}} \end{bmatrix} \begin{bmatrix} \mathbf{U}^* & \mathbf{0} \\ \mathbf{0} & \mathbf{I} \end{bmatrix} \begin{bmatrix} \mathbf{X} \\ \boldsymbol{\Upsilon} \end{bmatrix} \right\|_F^2 \tag{3.52}$$

By defining the first observable $g_1(\boldsymbol{x}) = \boldsymbol{x}$ as in (3.50), it is possible to retain the state of interest and still benefit from the DMD extension by a careful choosing of the remaining non-linear observables.

In the methodology defined in [47], functions g may not only represent non-linear functions of the state but physical measurements as well, such as outputs. In the case of a high dimensional system, the number of outputs l might be significantly smaller than the number of states n. Therefore, there may not be the need for SVD and projection to the low order subspace. The relation in (3.52) can be changed to accommodate physical measurements, by defining what are termed *deterministic states*: $X_d \in \mathbb{R}^d$. Assembling the snapshot matrices of such variables and inserting them into (3.52), the least squares problem boils down to

$$\min_{\begin{bmatrix}\tilde{\mathbf{A}} & \tilde{\mathbf{B}} \\ \tilde{\mathbf{C}} & \tilde{\mathbf{D}}\end{bmatrix}} \left\| \begin{bmatrix} \mathbf{X}'_d \\ \mathbf{X}' \\ \mathbf{Y} \end{bmatrix} - \begin{bmatrix} \mathbf{I} & \mathbf{0} & \mathbf{0} \\ \mathbf{0} & \mathbf{U} & \mathbf{0} \\ \mathbf{0} & \mathbf{0} & \mathbf{I} \end{bmatrix} \begin{bmatrix} \tilde{\mathbf{A}}_d & \mathbf{A}_{nd} & \tilde{\mathbf{B}}_d \\ \mathbf{A}_d & \tilde{\mathbf{A}}_{nd} & \tilde{\mathbf{B}}_{nd} \\ \tilde{\mathbf{C}}_d & \tilde{\mathbf{C}}_{nd} & \tilde{\mathbf{D}} \end{bmatrix} \begin{bmatrix} \mathbf{I} & \mathbf{0} & \mathbf{0} \\ \mathbf{0} & \mathbf{U}^* & \mathbf{0} \\ \mathbf{0} & \mathbf{0} & \mathbf{I} \end{bmatrix} \begin{bmatrix} \mathbf{X}_d \\ \mathbf{X} \\ \boldsymbol{\Upsilon} \end{bmatrix} \right\|_F^2 \tag{3.53}$$

where the subscript d refers to a term coupled with a deterministic state and the subscript nd refers to a term coupled with a nondeterministic state. The reduced order model is then computed by least squares:

$$\Theta_{\text{EIODMD}} = \begin{bmatrix} \mathbf{X}'_d \\ \mathbf{U}^*\mathbf{X}' \\ \mathbf{Y} \end{bmatrix} \begin{bmatrix} \mathbf{X}_d \\ \boldsymbol{\Sigma}\mathbf{V} \\ \boldsymbol{\Upsilon} \end{bmatrix}^{\dagger} \tag{3.54}$$

The reduced order matrices are then:

$$\tilde{\mathbf{A}} = \begin{bmatrix} \tilde{\mathbf{A}}_d & \mathbf{A}_{nd} \\ \mathbf{A}_d & \tilde{\mathbf{A}}_{nd} \end{bmatrix}, \quad \tilde{\mathbf{B}} = \begin{bmatrix} \tilde{\mathbf{B}}_d \\ \tilde{\mathbf{B}}_{nd} \end{bmatrix}, \quad \tilde{\mathbf{C}} = \begin{bmatrix} \tilde{\mathbf{C}}_d & \tilde{\mathbf{C}}_{nd} \end{bmatrix}, \quad \tilde{\mathbf{D}} = \tilde{\mathbf{D}} \tag{3.55}$$

where $\tilde{\mathbf{A}} \in \mathbb{R}^{(d+r)\times(d+r)}$, $\tilde{\mathbf{B}} \in \mathbb{R}^{(d+r)\times q}$, $\tilde{\mathbf{C}} \in \mathbb{R}^{l\times(d+r)}$ and $\tilde{\mathbf{D}} \in \mathbb{R}^{l\times q}$.

This extension of IODMD attempts to increase state reconstruction accuracy: The generalization of DMD to include non-linear observables, having as a starting point an analysis of the non-linear terms of the governing non-linear equations, attempts to improve the embedding of the non-linear dynamics in the linear model, thus improving the predictability of the model's states.

The code snippet Listing 3.2 is a MATLAB implementation of the EIODMD algorithm and illustrates how the code implementation in Listing 3.1 could be adapted to include the so-called deterministic states. Note that the deterministic states are a possible route to attempt finding a better set of coordinates. In this case, the states matrix may be augmented to include other variables, as compared to the states matrix in Listing 3.1.

Listing 3.2 Implementation of Extended Input-Output Dynamic Mode Decomposition in MATLAB

```
function [ROM]=eiodmd(states,Inputs,Outputs,Deterministic)

    %% INPUT ARGUMENTS
    %states:   states to build DMD matrices
    %inputs:   [U1 U2 U3 ... Un];
    %outputs:  [Y1 Y2 Y3 .... Yn];
    %deterministic: [Xd1 Xd2 Xd3 Xd4 ... Xdn];

    %define necessary matrices for IODMD
    X       = states(:,1:end-1);        %snapshot matrix
    X_p     = states(:,2:end);          %time-shifted snapshot matrix

    out     = Outputs(:,1:end-1);     %output data matrix
    out_p   = Outputs(:,2:end);       %ime shifted output data matrix

    inp     = Inputs(:,1:end-1);       %input data matrix
    inp_p   = Inputs(:,2:end);         %time shifted input data matrix

    Xd      = Deterministic(:,1:end-1); %deterministic data matrix
    Xd_p    = Deterministic(:,2:end);   %time shifted deterministic matrix

    %define truncation value for the SVD
    r=100

    %Perform SVD of X
    [U,S,V]=svds(X,r);

    %Use EIODMD solution
    matr=[ Xd_p; U'*X_p;out]*pinv([Xd; S*V';inp]);

    %Fetch A,B,C,D matrices
    A=matr(1:size(Xd,1)+r,1:size(Xd,1)+r);
    B=matr(1:size(Xd,1)+r,size(Xd,1)+r+1:end);
    C=matr(size(Xd,1)+r+1:end, 1:size(Xd,1)+r);
    D=matr(size(Xd,1)+r+1:end, size(Xd,1)+r+1:end);

    %compute state space Reduced Order Model
    ROM=ss(A,B,C,D,2);

end
```

3.7 Regularization

DMD-based identification methods, such as IODMD and its extended version, do not guarantee that its resulting models are stable. To enforce stability, an additional post-processing step is required. In [49] it is shown that directly solving a non-smooth constrained optimization problem can be a viable approach for stabilizing IODMD-derived systems while retaining the salient properties for approximating the output response. This is achieved by using an open-source MATLAB code, GRANSO (Gradient-based Algorithm Non-smooth Optimization).

A simpler, although model dependent post-processing tool has been developed and explored in the context of this work, based on the ideas in [50]. In the former, stability of an estimated system—from least squares and a finite number of data points—is imposed by adding a regularization term to the least squares cost function. Given the least squares problem to identify the system matrices $(\hat{\mathbf{A}}, \hat{\mathbf{B}}, \hat{\mathbf{C}}, \hat{\mathbf{D}})$:

$$\min_{\hat{\mathbf{A}},\hat{\mathbf{B}},\hat{\mathbf{C}},\hat{\mathbf{D}}} \left\| \begin{bmatrix} \hat{\mathbf{X}}_{i+1} \\ \mathbf{Y}_{i|i} \end{bmatrix} - \begin{bmatrix} \hat{\mathbf{A}} & \hat{\mathbf{B}} \\ \hat{\mathbf{C}} & \hat{\mathbf{D}} \end{bmatrix} \begin{bmatrix} \hat{\mathbf{X}}_i \\ \mathbf{U}_{i|i} \end{bmatrix} \right\|_F^2 \tag{3.56}$$

its solution can be presented as two solutions for independent least squares problems:

$$\begin{aligned} \begin{bmatrix} \hat{\mathbf{A}} & \hat{\mathbf{B}} \end{bmatrix} &= \hat{\mathbf{X}}_{i+1} \begin{bmatrix} \hat{\mathbf{X}}_i^T \mathbf{U}_{i|i}^T \end{bmatrix} \hat{\boldsymbol{\Sigma}}_{XU}^{-1} \\ \begin{bmatrix} \hat{\mathbf{C}} & \hat{\mathbf{D}} \end{bmatrix} &= \mathbf{Y}_{i|i} \begin{bmatrix} \hat{\mathbf{X}}_i^T \mathbf{U}_{i|i}^T \end{bmatrix} \hat{\boldsymbol{\Sigma}}_{XU}^{-1} \end{aligned} \tag{3.57}$$

where

$$\hat{\boldsymbol{\Sigma}}_{XU} = \begin{bmatrix} \hat{\mathbf{X}}_i \hat{\mathbf{X}}_i^T & \hat{\mathbf{X}}_i \mathbf{U}_{i|i}^T \\ \mathbf{U}_{i|i} \hat{\mathbf{X}}_i^T & \mathbf{U}_{i|i} \mathbf{U}_{i|i}^T \end{bmatrix} \tag{3.58}$$

Stability of the model is then imposed by adding a regularization term to the cost function from which (3.57) is determined. For a full explanation on the cost function formulation and its rationale, the reader is referred to [50]. The optimal solution is then given by

$$\begin{aligned} \begin{bmatrix} \tilde{\mathbf{A}}_c & \tilde{\mathbf{B}}_c \end{bmatrix} &= \begin{bmatrix} \hat{\mathbf{X}}_{i+1} \hat{\mathbf{X}}_i^T & \hat{\mathbf{X}}_{i+1} \mathbf{U}_{i|i}^T \end{bmatrix} \left(\hat{\boldsymbol{\Sigma}}_{XU} + c\mathbf{W}_e^{-1} \right) \\ &= \begin{bmatrix} \hat{\mathbf{A}} & \hat{\mathbf{B}} \end{bmatrix} \hat{\boldsymbol{\Sigma}}_{XU} \left(\hat{\boldsymbol{\Sigma}}_{XU} + c\mathbf{W}_e \right)^{-1} \end{aligned} \tag{3.59}$$

where $\hat{\boldsymbol{\Sigma}}_{XU}$ is defined in (3.58) and $\mathbf{W}_e$ as

$$\mathbf{W}_e = \begin{bmatrix} \mathbf{W} & \mathbf{0} \\ \mathbf{0} & \mathbf{0} \end{bmatrix} \tag{3.60}$$

The amount of regularization is characterized by the positive real scalar c and by the positive semidefinite[14] matrix $\mathbf{W}$. The same principle applied to subspace identification methods can be used in the IODMD models derivations. The analogous regularized solution in (3.59), in the context of IODMD, using the full explicit formulation in (3.41), is the following:

$$\begin{bmatrix} \tilde{\mathbf{A}}_c & \tilde{\mathbf{B}}_c \\ \tilde{\mathbf{C}}_c & \tilde{\mathbf{D}}_c \end{bmatrix} = \begin{bmatrix} \tilde{\mathbf{A}} & \tilde{\mathbf{B}} \\ \tilde{\mathbf{C}} & \tilde{\mathbf{D}} \end{bmatrix} \begin{bmatrix} \boldsymbol{\Sigma}\mathbf{V}\boldsymbol{\Sigma}^T\mathbf{V}^T & \boldsymbol{\Sigma}\mathbf{V}\boldsymbol{\Upsilon}^T \\ \boldsymbol{\Upsilon}\boldsymbol{\Sigma}^T\mathbf{V}^T & \boldsymbol{\Upsilon}\boldsymbol{\Upsilon}^T \end{bmatrix} \left(\begin{bmatrix} \boldsymbol{\Sigma}\mathbf{V}\boldsymbol{\Sigma}^T\mathbf{V}^T & \boldsymbol{\Sigma}\mathbf{V}\boldsymbol{\Upsilon}^T \\ \boldsymbol{\Upsilon}\boldsymbol{\Sigma}^T\mathbf{V}^T & \boldsymbol{\Upsilon}\boldsymbol{\Upsilon}^T \end{bmatrix} + c \begin{bmatrix} \mathbf{W} & \mathbf{0} \\ \mathbf{0} & \mathbf{0} \end{bmatrix} \right)^{-1} \tag{3.61}$$

[14] A positive semidefinite matrix is a symmetric matrix with the property that it never produces a negative value when any vector is multiplied by it in a specific way. More formally, a matrix $\mathbf{A}$ is positive semidefinite if, for any vector $\boldsymbol{x}$, the quadratic form $\boldsymbol{x}^T \mathbf{A} \boldsymbol{x} \geq 0$.

Matrices $\tilde{\mathbf{A}}_c$, $\tilde{\mathbf{B}}_c$, $\tilde{\mathbf{C}}_c$ and $\tilde{\mathbf{D}}_c$ correspond to the state space matrices after the post-processing step.

Although the least squares solution might be a stable one, in the event that it is not, using (3.61) ensures IODMD and EIODMD stability whilst having little or no effect in the outputs and states, as it will be seen in Chap. 7. In the latter chapter, regularization methods are discussed; however in this section the focus is to derive the theoretical mathematical post-processing based on the EIODMD formulation.

Listing 3.3 shows an example of implementing the regularization post-processing in a certain IODMD model. In the specific example, a matrix **M** is first generated and set to the identity matrix. It is possible to purposely modify how the location of each individual pole changes when applying this post-processing technique by choosing different entries in the **M** matrix. In the code snippet in Listing 3.3, by setting the first entry 100 times higher than the remaining ones, the change in the location of the lowest frequency pole should be higher. It is up to the user to make this choice and explore according to the final model application.

Listing 3.3 Implementation of regularization as a post-processing step in MATLAB

```
function [newROM]=regulatization(U,S,V,Outputs,Inputs,states,ROM,r)

    %define parameter c
    c=10;

    out        = Outputs(:,1:end-1);
    inp        = Inputs(:,1:end-1);

    %matrix sigma for IODMD computation
    kapa=[S*V';inp]*[S*V';inp]';

    %define matrix W (in this example it is identity)
    W=eye(size(ROM.A));

    %example of penalising one direction differently
    W(1,1)=100;

    %build auxiliary matrix for the regularisation, see equation (3.61)
    WW=[W, zeros(size(ROM.A,1),size(ROM.B,2));...
    zeros(size(ROM.B,2),size(ROM.A,1)), ...
    zeros(size(ROM.B,2) )];

    %apply regularization to state space matrices of ROM
    nrs=[ROM.A ROM.B;...
    ROM.C ROM.D]*kapa*(kapa+c*WW)^(-1);

    %Fetch new state space matrices after regularization
    An=nrs(1:r,1:r);
    Bn=nrs(1:r,r+1:end);
    Cn=nrs(r+1:end,1:r);
    Dn=nrs(r+1:end,r+1:end);

    %declare new state space ROM
    newROM=ss(An,Bn,Cn,Dn,2);

    %check for stability
    isstable(newROM)

end
```

This chapter started by focusing on the existing wind farm and turbine models often used for the purposes of wind farm flow control. Different steady-state models were introduced, as well as validation exercises of these. The dynamics governing the motion of a fluid are also shown as a means to emphasize their complexity and look for other approaches to modelling wind farms which avoid dealing directly with the existing equations. The alternative introduced was to merely use data, and from that compute models that embed to some degree the dynamics in the governing equations. The chapter proceeds to introduce the core method which will be used throughout the book, dynamic mode decomposition. The various variants of the algorithm are then introduced, namely input-output dynamic mode decomposition, so that a model which maps inputs to outputs can be identified. The Koopman operator theory was also introduced, as this provides relevant ideas to better capture the existing non-linear dynamics in the simplified reduced order model. The variant of the algorithm motivated by the Koopman operator theory, extended input-output dynamic mode decomposition, is also introduced. In the next chapter, the overarching systematic methodology to model wind farm flow control strategies is presented.

References

1. S. Boersma et al., A tutorial on control-oriented modeling and control of wind farms, in *2017 American Control Conference (ACC)* (2017), pp. 1–18
2. P.M.O. Gebraad, Data-Driven Wind Plant Control. PhD Thesis. Technische Universiteit Delft, 2014
3. NREL, *FLORIS. Version 2.4* (2021). https://github.com/NREL/floris
4. N.O. Jensen, F. Risø, *A Note on Wind Generator Interaction*. RISOE-M-2411. Risø National Laboratory (1983). ISBN: 978-87-55009-71-4. https://books.google.nl/books?id=5w3tvgEACAAJ
5. M. Bastankhah, F. Porteé-Agel, Experimental and theoretical study of wind turbine wakes in yawed conditions. J. Fluid Mech. **806**, 506–541 (2016)
6. M. Bastankhah, F. Porté-Agel, A new analytical model for wind-turbine wakes. Renew. Energy **70** (2014). Special issue on aerodynamics of offshore wind energy systems and wakes, pp. 116–123. ISSN: 0960-1481. https://doi.org/10.1016/j.renene.2014.01.002. https://www.sciencedirect.com/science/article/pii/S0960148114000317
7. L.A. Martınez-Tossas et al., The aerodynamics of the curled wake: a simplified model in view of flow control. Wind Energy Sci. **4**(1), 127–138 (2019). https://doi.org/10.5194/wes-4-127-2019. https://wes.copernicus.org/articles/4/127/2019/
8. N.G. Nygaard et al., TurbOPark: A Turbulence Optimised Park model. Technical Report, Ørsted (2022)
9. C.J. Bay et al., Addressing deep array effects and impacts to wake steering with the cumulative-curl wake model. Wind Energy Sci. Discuss. **2022**, 1–28 (2022). https://doi.org/10.5194/wes-2022-17. https://wes.copernicus.org/preprints/wes-2022-17/
10. B.M. Doekemeijer, E. Simley, P. Fleming, Comparison of the Gaussian wind farm model with historical data of three offshore wind farms. Energies **15**(6), 1964 (2022). ISSN: 1996-1073. https://doi.org/10.3390/en15061964. https://www.mdpi.com/1996-1073/15/6/1964
11. N.G. Nygaard et al., Large-scale benchmarking of wake models for offshore wind farms. J. Phys. Conf. Ser. **2265**(2), 022008 (2022). https://doi.org/10.1088/1742-6596/2265/2/022008

12. F.M.White, *Fluid Mechanics*, 6th edn. (McGraw-Hill, New York, 2008)
13. W. Munters, J. Meyers, An optimal control framework for dynamic induction control of wind farms and their interaction with the atmospheric boundary layer. English. Philosoph. Trans. R. Soc. London Ser. A Math. Phys. Eng. Sci. **375**(2091), 20160100 (2017). ISSN: 1364-503X
14. J. Annoni, Modeling for Wind Farm Control. PhD Thesis. University of Minnesota, 2016
15. P. Holmes et al., *Turbulence, Coherent Structures, Dynamical Systems and Symmetry*. English, 2nd edn. (Cambridge University Press, New York, 2002)
16. P. Schmid, J. Sesterhenn, Dynamic mode decomposition of numerical and experimental data, in *vol. 61st Annual Meeting of the APS Division of Fluid Dynamics* (American Physical Society, College Park, 2008)
17. N. Hamilton, M. Tutkun, R.B. Cal, Low-order representations of the canonical wind turbine array boundary layer via double proper orthogonal decomposition. Phys. Fluids **28**(2), 025103 (2016). https://doi.org/10.1063/1.4940659
18. D. Bastine et al., Towards a simplified dynamic wake model using POD analysis. Energies **8**(2), 895–920 (2015). ISSN: 1996-1073. https://doi.org/10.3390/en8020895. https://www.mdpi.com/1996-1073/8/2/895
19. M. Debnath et al., Towards reduced order modelling for predicting the dynamics of coherent vorticity structures within wind turbine wakes. Philosoph. Trans. R. Soc. A Math. Phys. Eng. Sci. **375**, 20160108 (2017). https://doi.org/10.1098/rsta.2016.0108
20. N. Hamilton et al., A generalized framework for reduced-order modeling of a wind turbine wake. Wind Energy **21**, 373–390 (2018). https://doi.org/10.1002/we.2167
21. J.N. Kutz et al., *Dynamic Mode Decomposition: Data-Driven Modelling of Complex Systems* (SIAMR, 2016)
22. R.S.S. Mukherjee, Big data–concepts, applications, challenges and future scope. Int. J. Adv. Res. Comput. Commun. Eng. **5**, 66–74 (2016)
23. P. Schmid, Dynamic mode decomposition of numerical and experimental data. J. Fluid Mech. **656**, 5–28 (2010)
24. P. Schmid, K.E. Meyer, O. Pust, Dynamic mode decomposition and proper orthogonal decomposition of flow in a lid-driven cylindrical cavity, in *8th International Symposium on Particle image Velocimetry - PIV09* (2009)
25. J.H. Tu et al., On dynamic mode decomposition: theory and applications. J. Comput. Dyn. **1**(2), 391–421 (2014). ISSN: 2158-2505
26. J. Proctor, P. Eckhoff, Discovering dynamic patterns from infectious disease data using dynamic mode decomposition, in *International Health* (2015)
27. B. Brunton, L. Johnson, N.K.J. Ojemann, Extracting spatial-temporal coherent patterns in large-scale neural recordings using dynamic mode decomposition. J. Neurosci. Methods **258**, 1–15 (2016). ISSN: 0165-0270
28. G.V. Iungo et al., Data-driven reduced order model for prediction of wind turbine wakes. J. Phys. Conf. Ser. **625**, 012009 (2015). https://doi.org/10.1088/1742-6596/625/1/012009
29. J. Annoni, P.M.O. Gebraad, P. Seiler, Wind farm flow modeling using an input-output reduced-order model, in *2016 American Control Conference (ACC)* (2016), pp. 506–512
30. N. Cassamo, J.-W. van Wingerden, Model predictive control for wake redirection in wind farms: A koopman dynamic mode decomposition approach, in *2021 American Control Conference (ACC)* (2021), pp. 1776–1782. https://doi.org/10.23919/ACC50511.2021.9482631
31. J. Liew et al., Streaming dynamic mode decomposition for short-term forecasting in wind farms. English. Wind Energy **25**, 719–734 (2022). ISSN: 1095-4244. https://doi.org/10.1002/we.2694
32. G.H. Golub, C.F. Van Loan, *Matrix Computations*, 3rd edn. (The Johns Hopkins University Press, Baltimore, 1996)
33. J.L. Proctor, S.L. Brunton, J.N. Kutz, Dynamic mode decomposition with control. SIAM J. Appl. Dyn. Syst. **15**(1), 142–161 (2016)
34. J. Annoni, P. Seiler, A low-order model for wind farm control, in *2015 American Control Conference (ACC)* (2015), pp. 1721–1727

35. M. Viberg, Subspace-based methods for the identificaiton of linear time-invariant systems. Automatica **31**(12), 1835–1851 (1995)
36. C. Rowley et al., Spectral analysis of nonlinear flows. J. Fluid Mech. **641**, 115–127 (2009)
37. B.O. Koopman, Hamiltonian systems and transformation in hilbert space. Proc. Nat. Acad. Sci. **17**(5), 315–318 (1931)
38. S. Brunton et al., Koopman observable subspaces and finite linear representations of nonlinear dynamical systems for control. PloS One **11** (2015). https://doi.org/10.1371/journal.pone.0150171
39. I. Mezic, Analysis of fluid flows via spectral properties of the Koopman operator. Ann. Rev. Fluid Mech. **45**, 357–378 (2013)
40. M. Williams, I. Kevrekidis, C. Rowley, A data-driven approximation of the Koopman operator: extending dynamic mode decomposition. J. Nonlinear Sci. **25**, 1307–1346 (2014)
41. I. Kevrekidis, C. Rowley, M. Williams, A kernel-based method for data-driven Koopman spectral analysis. J. Comput. Dyn. **2**, 247–265 (2016)
42. S.L. Brunton et al., Koopman invariant subspaces and finite linear representations of nonlinear dynamical systems for control. PLOS **11**(2), e0150171 (2016). Ed. by Hans AEditor Kestler
43. J. Kutz, J. Proctor, S. Brunton, Applied Koopman theory for partial differential equations and data-driven modeling of spatio-temporal systems. Complexity **2018**, 1–16 (2018)
44. H. Lu, D.M. Tartakovsky, Predictive accuracy of dynamic mode decomposition. In: arXiv: Numerical Analysis (2019)
45. J. Proctor, S. Brunton, J. Kutz, Generalizing Koopman theory to allow for inputs and control. SIAM J. Appl. Dyn. Syst. **17**, 909–930 (2016)
46. M. Korda, I. Mezić, Linear predictors for nonlinear dynamical systems: Koopman operator meets model predictive control. Automatica **93**, 149–160 (2018). ISSN: 0005-1098
47. H. Arbabi, M. Korda, I. Mezic, A data-driven Koopman model predictive control framework for nonlinear partial differential equations, in *2018 IEEE Conference on Decision and Control (CDC)* (2018), pp. 6409–6414
48. N. Cassamo, J.-W. van Wingerden, On the potential of reduced order models for wind farm control: a Koopman dynamic mode decomposition approach. Energies **13**(24), 6513 (2020). ISSN: 1996-1073
49. P. Benner, C. Himpe, T. Mitchell, On input-output dynamic mode decomposition. Adv. Comput. Math. **44**, 1751 (2018)
50. T. Van Gestel et al., Identification of stable models in subspace identification by using regularization. IEEE Trans. Autom. Control **46**, 1416–1420 (2001)

Chapter 4
Methodology for Modelling Wind Farm Control Strategies

In this chapter, the main steps for finding models for wind farms where control strategies at the plant level are tested using IODMD and EIODMD are shown. The methodology follows the system identification loop, a classical procedure used for model identification and validation. The procedure applied for IODMD and EIODMD is fully detailed in this chapter, so that in the upcoming chapters these can be readily applied to practical data from simulations to derive models.

4.1 The System Identification Loop

In this book, the focus is to compute and explore Reduced Order Models of wind farm control strategies. For such purpose, a systematic and structured approach is followed, often referenced in the controls literature as the system identification loop. This approach is used in the following chapters to derive models for both Wake Redirection Control and Axial Induction Control.

The system of interest in this work is a small wind farm, where two turbines are placed apart, and it is schematically represented in Fig. 4.1.

Several variables within the wind farm are subject to variations, such as the power output of each turbine, the loads on each turbine and the wind characteristics at different spatial points, amongst others. The model aims to relate and explain such variations. The directly manipulated variable—the input—will be the yaw angle for the case of Wake Redirection Control and the collective blade pitch angle for the Axial Induction Control—and the outputs the generator powers of both turbines.

Various routes exist for the task of model building. For the purpose of this book the route of *system identification* is used. This methodology is focused on experimentation, where input and output signals from the system are recorded and subjected to data analysis so as to infer a model. System Identification is a methodology by which one can generate a set of models based on the observed

N. Cassamo, J.-W. van Wingerden, *Data-driven Modelling of Wind Farm Flow Control Strategies*, https://doi.org/10.1007/978-3-031-84116-3_4

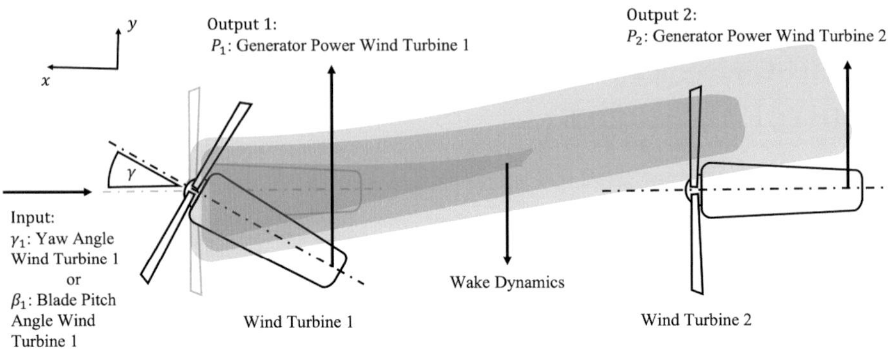

Fig. 4.1 Wind farm system with one input, which could be the upstream turbine yaw angle or the blades pitch angle and two outputs—generator power of both turbines

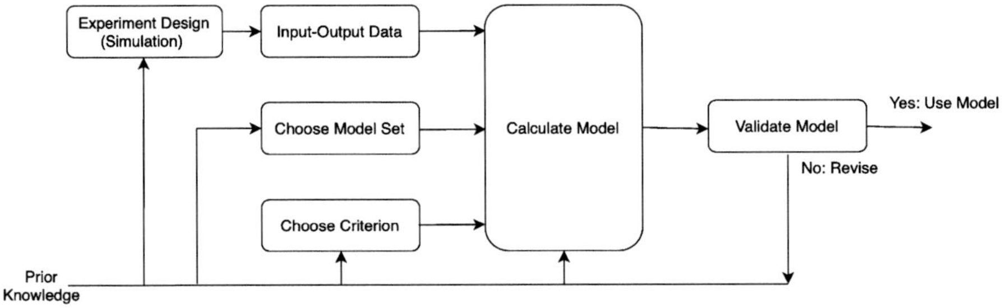

Fig. 4.2 The system identification loop

input-output data from the system and choose the best model(s) in accordance with a defined criterion. Therefore, the problem of system identification is specified by three elements [1, 2]:

- A dataset $\mathscr{D}$ obtained by input-output measurements
- A set of candidate models $\mathscr{M}$
- A criterion $\mathscr{L}$ to evaluate candidate models and select the best model(s)

A flow diagram of the typical system identification procedure adopted is represented in Fig. 4.2. For each step of the process, a further breakdown is provided, and the tools in the following chapters are introduced and fully explained.

4.1.1 Simulation Design

The input-output dataset $\mathscr{D}$ is obtained by performing a CFD simulation in SOWFA. SOWFA is a Large Eddy Simulation (LES) of the three dimensional wind flow around one or more turbine rotors in the Atmospheric Boundary Layer (ABL)[1] developed by the NREL, United States. In the LES method, in order to limit the computational cost of the simulation, the larger scales of the flow field are resolved by solving the temporally and spatially unsteady Navier-Stokes equations (with additional stress terms that empirically model the effect of the turbulence in the smaller unresolved scales), as described in [3]. The rotating rotor blades are modelled through an actuator line approach. The actuator lines are coupled with the FAST turbine aeroelastic simulator tool which calculates the loads, power and rotor speed of each turbine. SOWFA is a widely used tool in the literature to obtain accurate data. More information can be found in [4].

In the work here presented, wind with a uniform inflow profile[2] is used to demonstrate the working principles of DMD. The case of a wind farm with two turbines is studied, to assess the gain in energy in the second turbine. The second turbine is situated 5 rotor diameters (5D) behind the upstream turbine. The turbine used for the simulation is the DTU 10MW turbine, with a rotor diameter of 178.3 m. Detailed design characteristics of the turbine can be found in [5].

The full domain and the characteristics of the inflow used for the CFD simulation are the same as the ones used in [6]. The complete domain of the wind farm used is of 2.5 km $\times$ 1 km $\times$ 0.6 km. The uniform inflow speed is of 9 m/s. The ABL stability is neutral, and the inflow was generated through a so-called precursor simulation. The surface roughness and turbulence intensities were set to zero for this simulation. Note that these conditions are not representative of true realistic conditions found in offshore wind farms, where undisturbed inflow is turbulent. However, these simplified conditions provide a dataset which serves as a starting point. Further investigations with more complex conditions can be performed afterwards in light of the results from the simplified conditions.

The domain size used for IODMD has dimensions of approximately 1.71 km in the x direction (lengthwise), 343 km in the y direction and 300 km in the z direction. The domain is represented in Fig. 4.3. The dimensions have been normalized by the rotor diameter. A total of 550 points in x, 111 points in y and 99 points in z are sampled during the simulation, which means the complete flow field is solved in each of these points. Thus, a total of 6 million points, each one containing information on the (u, v, w) velocity components, is taken.

Due to the large volume of information generated during the simulation, a pre-processing step of the data is performed in a computer cluster, aiming to retrieve only information at each fourth point. This decreases the total number of points from 6 million to roughly 100 thousand samples, decreasing computational effort. Flow field information is sampled at each 2 seconds, which corresponds to a sampling frequency of 0.5 Hz. According to [7, 8], where AIC is analysed, the majority of

[1] The lowest region of the atmosphere where the Earth's surface strongly influences temperature and wind through the turbulent transfer of air mass.

[2] A specific case characterized by wind conditions with the same speed and direction across the rotor swept area. No turbulence, wind gusts or other more complex disturbances that could affect turbine performance are considered.

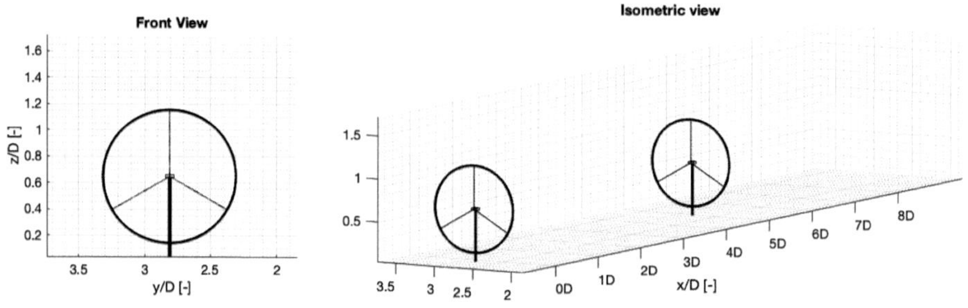

Fig. 4.3 Two turbine set layout in SOWFA. On the left, the front view of upstream turbine is shown. On the right, a three dimensional perspective is visible, depicting the distance of 5 rotor diameters (890 m) between upstream and downstream turbine

the frequency content within a wake 4 rotor diameters downstream of the upstream turbine has a lower frequency than 0.5 Hz, concentrating around 0.25 Hz. Higher frequencies are related to the presence of a shear layer. The decision of using a sampling period T_s of 2 seconds is also related with computational effort, as using a lower one would increase the total number of data points and using IODMD could become infeasible on a normal computer. In summary, the sampling period chosen is a trade-off in terms of capturing the highest frequencies and avoiding extreme computational costs, although it is believed the most relevant frequencies are captured and the Nyquist criterion[3] satisfied.

In addition to flow field, a variety of data on the turbines is also saved, more specifically information of the blades, nacelle, rotor, tower and generator. This information is sampled at each 0.2 seconds.

4.1.2 Model Selection

The candidate models are obtained in the well-known *state space representation*, often used in control engineering.

> The state space description provides the dynamics as a set of coupled first-order differential equations in a set of internal variables known as *state variables*, together with a set of algebraic equations that combine the state variables into physical output variables. The state variable is a physical quantity which, in the absence of external excitation, completely determines the evolution of the system.

[3] The Nyquist criterion states that a repetitive waveform can be correctly reconstructed provided that the sampling frequency is greater than double the highest frequency to be sampled.

The set of n coupled first-order differential equations, known as state equations, follows the general form:

$$\begin{aligned}\dot{x}_1 &= f_1(\boldsymbol{x}, \boldsymbol{u}, t)\\ \dot{x}_2 &= f_2(\boldsymbol{x}, \boldsymbol{u}, t)\\ \vdots &= \vdots\\ \dot{x}_n &= f_n(\boldsymbol{x}, \boldsymbol{u}, t)\end{aligned} \tag{4.1}$$

where $\dot{x}_i = dx_i/dt$ and each of the functions may be a general non-linear, time-varying function of the state variables $\boldsymbol{x}$ and the system input $\boldsymbol{u}$.

State equations are commonly written in vector form, with the set of n state variables written as a *state vector* $\boldsymbol{x}(t) = [x_1(t), x_2(t), \ldots, x_n(t)]^T$ and the set of q inputs written as an input vector $\boldsymbol{u}(t) = [u_1(t), u_2(t), \ldots, u_q(t)]^T$.

The models used in this work are Linear Time Invariant (LTI); that is the system is described by linear differential equations with constant coefficients. For an LTI system of order n and with q inputs, equations in (4.1) become a set of n first-order linear differential equations with constant coefficients:

$$\begin{aligned}\dot{x}_1 &= a_{11}x_1 + a_{12}x_2 + \cdots + a_{1n}x_n + b_{11}u1 + \cdots + b_{1r}u_q\\ \dot{x}_2 &= a_{21}x_1 + a_{22}x_2 + \cdots + a_{2n}x_n + b_{21}u1 + \cdots + b_{2r}u_q\\ \vdots &= \vdots\\ \dot{x}_n &= a_{n1}x_1 + a_{n2}x_2 + \cdots + a_{nn}x_n + b_{n1}u1 + \cdots + b_{nr}u_q\end{aligned} \tag{4.2}$$

where the coefficients a_{ij} and b_{ij} are constants that describe the system. This set of n equations defines the derivatives of the state variables to be a weighted sum of the state variables and the system inputs. Equation (4.2) may be written compactly in a matrix form:

$$\frac{d}{dt}\begin{bmatrix}x_1\\ x_2\\ \vdots\\ x_n\end{bmatrix} = \begin{bmatrix}a_{11} & a_{12} & \cdots & a_{1n}\\ a_{21} & a_{22} & \cdots & a_{2n}\\ \vdots & & \ddots & \vdots\\ a_{n1} & a_{n2} & \cdots & a_{nn}\end{bmatrix}\begin{bmatrix}x_1\\ x_2\\ \vdots\\ x_n\end{bmatrix} + \begin{bmatrix}b_{11} & \cdots & b_{1q}\\ b_{21} & \cdots & b_{2q}\\ \vdots & & \vdots\\ b_{n1} & \cdots & b_{nq}\end{bmatrix}\begin{bmatrix}u_1\\ \vdots\\ u_q\end{bmatrix} \tag{4.3}$$

which may be summarized as

$$\dot{\boldsymbol{x}} = \mathbf{A}\boldsymbol{x} + \mathbf{B}\boldsymbol{u} \tag{4.4}$$

where the state vector $\boldsymbol{x}$ is a column vector of length n, the input vector $\boldsymbol{u}$ is a column vector of length q, $\mathbf{A}$ is an $n \times n$ square matrix of the constant coefficients a_{ij} and $\mathbf{B}$ is an $n \times q$ matrix of the coefficients b_{ij} that weigh the inputs. This nomenclature is not new and has already been introduced in Chap. 3. Matrices $\mathbf{A}$ and $\mathbf{B}$ in Eq. (4.4) correspond to the reduced order model matrices $\tilde{\mathbf{A}}$ and $\tilde{\mathbf{A}}$ in Eq. (3.42). However in this chapter they have been defined agnostic to their application.

The system output, defined to be any variable of interest, can be represented by a linear combination of the state variables x_i and the system inputs u_i. Writing the equivalent equations in a compact form yields

$$\begin{bmatrix} y_1 \\ y_2 \\ \vdots \\ y_l \end{bmatrix} = \begin{bmatrix} c_{11} & c_{12} & \cdots & c_{1n} \\ c_{21} & c_{22} & \cdots & c_{2n} \\ \vdots & & \ddots & \vdots \\ c_{l1} & c_{l2} & \cdots & c_{ln} \end{bmatrix} \begin{bmatrix} x_1 \\ x_2 \\ \vdots \\ x_n \end{bmatrix} + \begin{bmatrix} d_{11} & \cdots & d_{1q} \\ d_{21} & \cdots & d_{2q} \\ \vdots & & \vdots \\ d_{n1} & \cdots & d_{lq} \end{bmatrix} \begin{bmatrix} u_1 \\ \vdots \\ u_q \end{bmatrix} \tag{4.5}$$

which can be further summarized in a compact notation:

$$\boldsymbol{y} = \mathbf{C}\boldsymbol{x} + \mathbf{D}\boldsymbol{u} \tag{4.6}$$

where $\boldsymbol{y}$ is a column vector of the output variables $y_i(t)$, $\mathbf{C}$ is an $l \times n$ matrix of the constant coefficients c_{ij} that weight the state variables and $\mathbf{D}$ is an $l \times q$ matrix of the constant coefficients d_{ij} that weight the system inputs. In discrete time, the differential equations become finite differences, and the complete system model for the LTI system is defined by

$$\begin{aligned} \boldsymbol{x}_{k+1} &= \mathbf{A}\boldsymbol{x}_k + \mathbf{B}\boldsymbol{u}_k \\ \boldsymbol{y}_k &= \mathbf{C}\boldsymbol{x}_k + \mathbf{D}\boldsymbol{u}_k \end{aligned} \tag{4.7}$$

Matrices **A** and **B** are properties of the system and are determined by the system structure and elements. The output equation matrices **C** and **D** are determined by the particular choice of output variables. All four matrices are computed using the algorithms described in Chap. 3.

4.1.3 Criterion of Fit

The fitness of the models is measured in two ways. The first concerns the evaluation of the models output. The predicted and measured outputs are compared and the degree with which they are associated quantified using the Variance Accounted For criteria. The second concerns the evaluation of the models state trajectory. The states of the models, i.e. the reconstructed wake, are compared with the wake from the simulation, and the differences are quantified using the average Normalized Root Mean Squared Error criteria.

The fitness of the model can be measured resorting to two indicators. First, it can be measured in its ability to map the inputs to the outputs, functioning as a good predictor of the generated power in both turbines as a function of the first turbine yaw or blade pitch angle. For this purpose, the criterion to use is Variance Accounted For (VAF), defined as [9],

$$\text{VAF}_i(y_i(k), \hat{y}_i(k,\theta)) = \left(1 - \frac{\frac{1}{N}\sum_{k=1}^{N} \mid y_i(k) - \hat{y}_i(k,\theta) \mid^2}{\frac{1}{N}\sum_{k=1}^{N} \mid y_i(k) \mid^2}\right) \cdot 100 \ [\%] \tag{4.8}$$

where $y_i(k)$ is the ith measured output at time instant k, $\hat{y}_i(k)$ the ith predicted output at time instant k, θ the model parameters and N the number of samples. The VAF criterion gives a quantitative measure of the discrepancy between the model and data. Higher percentages of VAF indicate a stronger strength of association.

Second, it can be measured in its ability to reconstruct the high order representation of the states $\boldsymbol{x}_k$, functioning as a good predictor of the wake dynamics as a function of the model input (e.g. upstream turbine yaw angle or blade pitch angle). For this purpose, the criterion to use is the Normalized Root Mean Squared Error (NRMSE), defined as

$$\begin{aligned} \text{NRMSE}_k(\boldsymbol{x}(k), \hat{\boldsymbol{x}}(k,\theta)) &= \frac{\text{RMSE}(\boldsymbol{x}(k), \hat{\boldsymbol{x}}(k,\theta))}{x_{\max}(k) - x_{\min}(k)} \\ &= \frac{\sqrt{\frac{\sum_{i=1}^{N}(x_i(k)-\hat{x}_i(k,\theta))^2}{N}}}{x_{\max}(k) - x_{\min}(k)} \cdot 100 \ [\%] \end{aligned} \tag{4.9}$$

where $\boldsymbol{x}(k)$ is the state vector at time instant k, $\hat{\boldsymbol{x}}(k)$ the predicted state vector at time instant k, θ the model parameters and N the number of samples, and the subscript i refers to the ith state in the state vector. The NRMSE is computed using the representation of the high order state vector at each time instant, and its mean value, considering the NRMSE from all time instants, $\overline{\text{NRMSE}}$ is taken to serve as a fit criteria.

4.1.4 Data Pre-processing

The data is pre-processed before being used to calculate candidate models. This process includes choosing only the most relevant information, resampling turbine information and removing trends and offsets from data.

1. *Use flow field information between the first and second turbines*: The domain size used for the simulation in SOWFA captures the flow information behind the

second turbine. Nevertheless, considering that the focus is on the wake dynamics between the first and second turbines—which explains power variations on the second—only this data is used to build the models. The reader is welcomed to attempt other spatial resampling strategies and assess their impact.

2. *Resample turbine information*: Velocity data is sampled at every 2 seconds, whereas turbine data is sampled at every 0.2 second. To compute the model, turbine data is resampled, so that both sampling frequencies are equal.

3. *Remove trends and offsets from input-output information*: A common procedure in systems identification is to remove offsets and trends from the data, so that the model focuses on the true intrinsic dynamics of the system. For this reason, turbine data used for input-output measurements is detrended by removing the mean value at steady-state operation.

The third rationale may be applied to the flow field information. In the POD methodology, it is common to subtract the mean value. However, subtracting a mean of the data reduces DMD to the temporal discrete Fourier transform, which may be restrictive and undesirable, as indicated in [10]. An alternative is to subtract an equilibrium point of the dynamics. This is often the case for fluid flows, where the equilibrium solution is a steady flow—commonly referred to as *base flow* $\boldsymbol{x}_b$. As the author in the previous article notes, subtracting the equilibrium appears to have significant advantages over using the mean.

In this work, and following the equations governing the dynamics of fluids as introduced in the previous chapter, three quantities are defined to distinguish the data to be used: $(\boldsymbol{u}, \boldsymbol{v}, \boldsymbol{w})_k$ refers to the velocity vector from the simulation at each time instant, $(\bar{\boldsymbol{u}}, \bar{\boldsymbol{v}}, \bar{\boldsymbol{w}})$ to the base flow and $(\boldsymbol{u}', \boldsymbol{v}', \boldsymbol{w}')_k$ to the difference between the velocity vector at each time instant and the base flow. The last may be interpreted as an unsteady flow, somewhat similar to the terms in the RANS equations.

4.1.5 Model Calculation

The input-output DMD variants are used to compute the wind farm state space ROM. As discussed in Sect. 3.6, the Koopman operator generalizes DMD to allow any observables to serve as a state. This decision process of choosing the states is discussed . Note that input-output information is maintained fixed, corresponding to the signals schematically represented in Fig. 4.1.

In order to establish a baseline result, IODMD is first used, where the streamwise velocity u is used for the states. This corresponds to the traditional approach in the literature [7, 11]. This methodology is dubbed IODMD_u for the remaining of the next sections. To evaluate the possible benefits of using alternatives to IODMD_u motivated by the Koopman operator to better reproduce the dynamics, three different pathways can be followed to calculate the models:

1. *Use any variable or combination of variables as states*: Instead of using u as a state, this is replaced by another variable, such as v or w or even a combination of all. This attempts to find the ones that are most indicative to recover the dynamics. The results of this first formulation are referred to as IODMD$_{xx}$, where xx is the observable used in replacement of u.

2. *Augment the state vector u by another flow variable*: This process allows to both maintain an interpretable model and gain insights into the physics of the problem whilst attempting to improve the accuracy of the model based on the concepts behind the Koopman operator theory. More specifically, reintroducing DMD notation, $g_1(x)$ is chosen to be u, and combinations with different $g_2(x)$ are experimented. This formulation is termed EIODMD$_{xx,yy}$, where xx is the first observable and yy the second observable used to augment the original snapshot data matrix.

3. *Augment the state vector u by any variable*: The third hypothesis focuses on augmenting the flow field information with turbine information, following the ideas laid out in [12].

To build the models, the SVD is taken by choosing a certain truncation level r. Singular values (and the corresponding right and left singular vectors) are continuously added to form state space models of increasing size. It follows that a total of r models are calculated, each one of increasing size, in order to find the optimal number of modes that best replicates the dynamics. A truncation r of the SVD equal to 100 is chosen. Empirical evidence in the literature, such as [11, 13], shows that even models with no more than 20 states can accurately reproduce the dynamics; hence 100 should be enough to compute satisfactory models and understand the impact of adding increasing states.

This chapter introduced the methodology to calculate and validate the reduced order models for the wind farm flow control strategies being studied in this book. The specifics of the simulations used to retrieve the data are mentioned, and the state space model is fully defined mathematically. The criteria used to evaluate the suitability of the models are also mentioned, and the different data-driven techniques to be used in the following chapter are discussed. In the next chapter, this methodology is applied to the first flow control case study by calculating, validating and investigating a reduced order model for wake redirection control.

References

1. L. Ljung, *System Identification: Theory for the User* (Prentice Hall PTR, Upper Saddle River, 1999)
2. T. Katayama, *Subspace Methods for System Identification* (Springer, Berlin, 2005)
3. P.M.O. Gebraad, Data-Driven Wind Plant Control. PhD Thesis. Technische Universiteit Delft, 2014

4. M.J. Churchfield et al., A numerical study of the effects of atmospheric and wake turbulence on wind turbine dynamics. J. Turbulence **13**, N14 (2012)
5. F. Zahle et al., *The DTU 10-MW Reference Wind Turbine*. Section for Aeroelastic Design and Section for Structures Technical University of Denmark DTU Wind Energy—Risø Campus (2013)
6. J.A. Frederik et al., The helix approach: using dynamic individual pitch control to enhance wake mixing in wind farms. Wind Energy **23**(8), 1739–1751 (2020)
7. J. Annoni, P.M.O. Gebraad, P. Seiler, Wind farm flow modeling using an input-output reduced-order model, in *2016 American Control Conference (ACC)* (2016), pp. 506–512
8. J. Annoni, Modeling for Wind Farm Control. PhD Thesis. University of Minnesota, 2016
9. M. Verhaegen, V. Verdult, N. Bergboer, *Filtering and System Identification: An Introduction to using Matlab Software*. Software manual for the lti system identification toolbox. Delft University of Technology, Delft Center for Systems and Control (2007)
10. K. Chen, J. Tu, C. Rowley, Variants of dynamic mode decomposition: boundary condition, Koopman, and Fourier analyses. J. Nonlinear Sci. **22**, 887–915 (2012)
11. G.V. Iungo et al., Data-driven reduced order model for prediction of wind turbine wakes. J. Phys. Conf. Ser. **625**, 012009 (2015). https://doi.org/10.1088/1742-6596/625/1/012009
12. H. Arbabi, M. Korda, I. Mezic, A data-driven Koopman model predictive control framework for nonlinear partial differential equations, in *2018 IEEE Conference on Decision and Control (CDC)* (2018), pp. 6409–6414
13. J. Annoni, P. Seiler, A low-order model for wind farm control, in *2015 American Control Conference (ACC)* (2015), pp. 1721–1727

Chapter 5
Data-Driven Modelling of Wake Redirection Control

In this chapter, the data driven methodologies introduced in Chap. 4 are used to derive a Reduced Order Model (ROM) for a set of two wind turbines where Wake Redirection Control is tested in a simulation environment using SOWFA. The contents of this chapter follow the various steps taken in the systems identification procedure already introduced. The dataset and code used in this chapter are available in https://github.com/nassircassamo/IODMD_SOWFA and [1, 2], and the reader is invited to explore and test it with the aim of making the learning process a hands-on and enjoyable experience.

5.1 Simulation and Data Record

A simulation of 2000 seconds (approximately 33 minutes of real time simulated) is performed where Randomized Binary Signals (RBSs)[1] provided as a setpoint for the yaw angle of the first turbine γ_{t1} serve as inputs.[2] The procedure is depicted in Fig. 5.1. The yaw angle is first set to $-10°$ and then varied to $-30°$ randomly throughout the simulation. The yaw of the second turbine γ_{t2} remains the same throughout the simulation.

The resulting generator power of both turbines during the simulation varies as a result of the yaw angle and wake deflection. The intention is to model these dynamics in order to accurately predict the power generated in each turbine, which is illustrated in Fig. 5.2.

In the first five minutes of the simulation, the first turbine yaw angle is set to $-10°$. This procedure has direct impact on the power produced in the downstream turbine: As the wake travels, a spike in power appears at 2 minutes and 30 seconds.

---

[1] A random process that can assume one of the two possible values at any time.

[2] Turbine information is saved in files automatically generated by SOWFA, which are then read using MATLAB functions developed by NREL.

N. Cassamo, J.-W. van Wingerden, *Data-driven Modelling of Wind Farm Flow Control Strategies*, https://doi.org/10.1007/978-3-031-84116-3_5

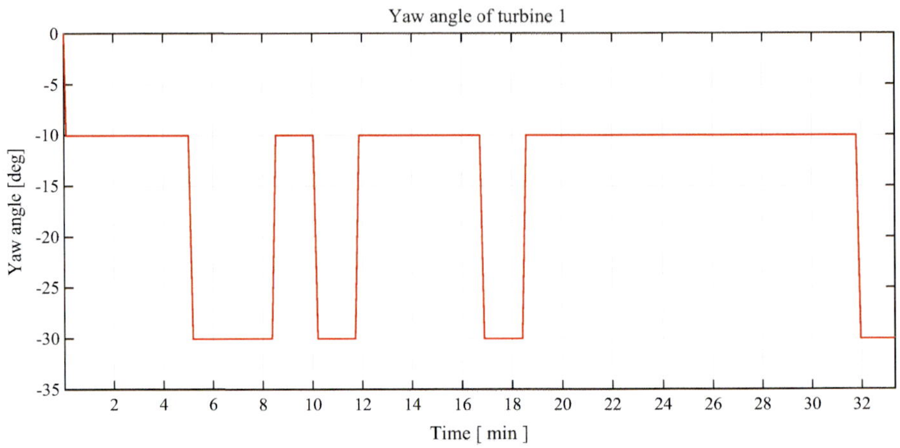

Fig. 5.1 Yaw angle variation of upstream turbine γ_{t1} during simulation. The yaw angle is first set to $-10°$ at the beginning of the simulation and then randomly varied to $-30°$

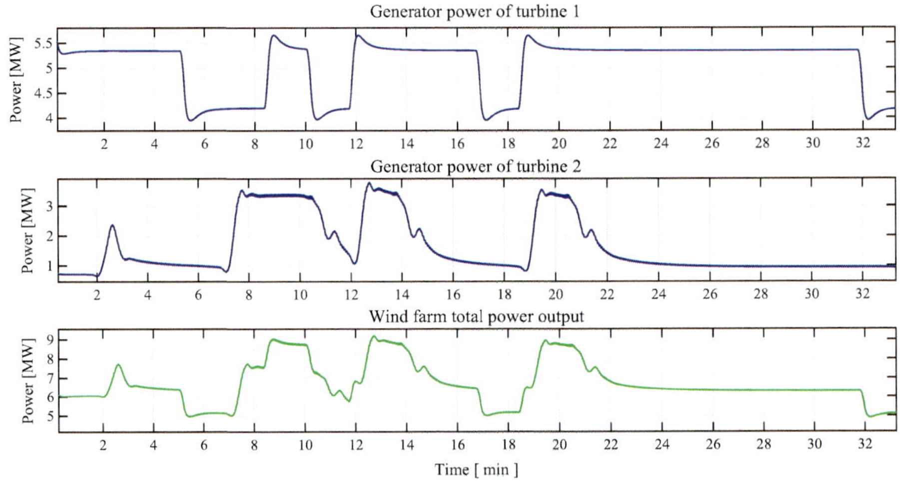

Fig. 5.2 Generator power variations of both turbines (in blue) during simulation and total generated power at wind farm level (in green)

This time interval serves the purpose of parameter setting and ensures that the wake is fully developed. The RBS used for systems identification starts acting after the initial 5 minutes.

The observed dynamics can be explained by inspecting the wake's behaviour as it propagates downstream. Figures 5.3 and 5.4[3] are reconstructions from the

[3] These snapshots were obtained using the function `wake_vorticity_deflection_yaw.m`. This function is available in https://github.com/nassircassamo/IODMD_SOWFA.

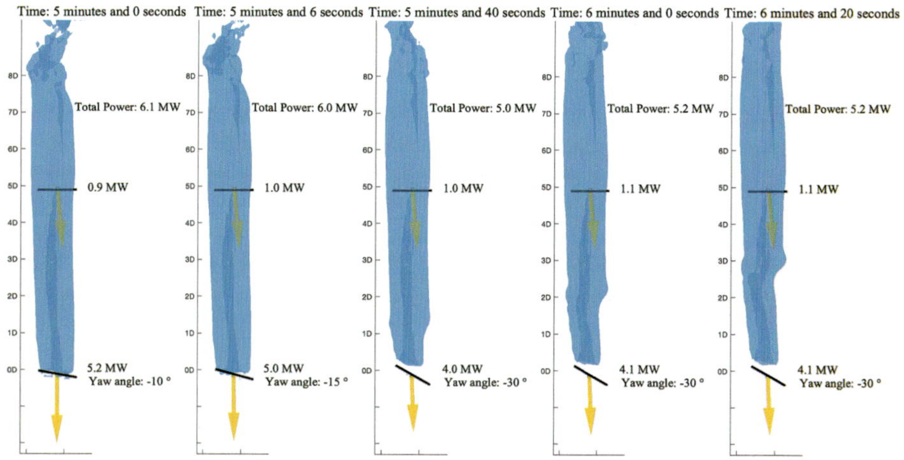

Fig. 5.3 Wake deflection dynamics: (1) Turbine starts yawing from $-10°$ and (2) continues to yaw until reaching fully yawed setting of $-30°$, (3) rotor is fully yawed and wake shows initial signs of deflection to the left, (4) wake starts propagating downstream and (5) wake continues to propagate, reaching a distance of 3 rotor diameters in more than 1 minute after being fully yawed

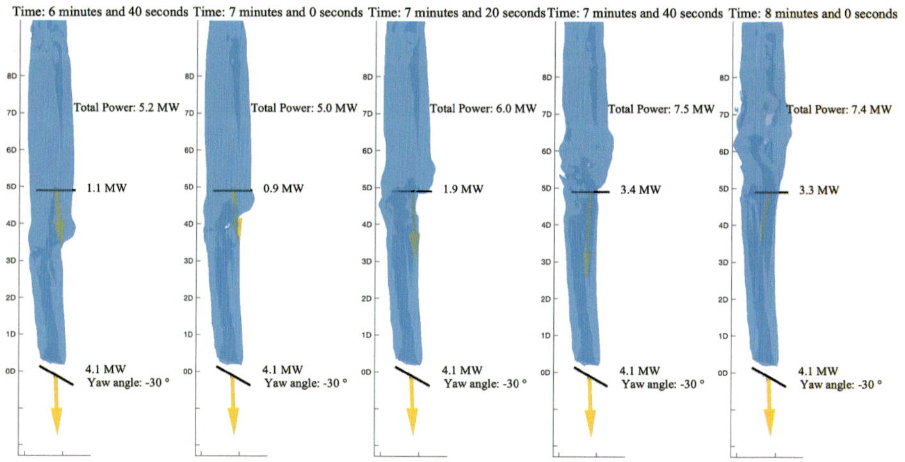

Fig. 5.4 Wake deflection dynamics: (1) and (2) Wake continues to propagate and deflect, (3) wake reaches downstream turbine rotor, and an increase of 1 [MW] is registered, (4) turbine 2 power output increases to 3.3 [MW] as deflected wake overlaps downstream rotor less and (5) wake continues to propagate behind turbine 2 and power output maintains a maximum of approximately 3.3 [MW]

simulation and allow to visually comprehend the relation between the wake's characteristics and the resulting generator power. The isosurface of the absolute vorticity field with value 5.65 is represented in blue, and it becomes clear how the deflection of the wake leads to increased power. Yellow arrows represent the thrust force direction.

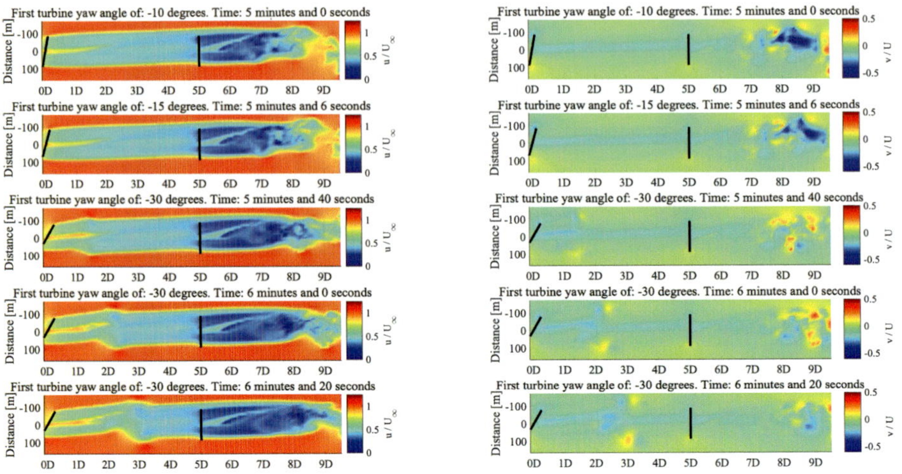

Fig. 5.5 Wake deflection dynamics (normalized streamwise and spanwise velocities, on the left and right sides, respectively): (1) and (2) First turbine starts yawing, (3) turbine is fully yawed $-30°$ and shows signs of initial deflection and (4) and (5) wake continues to propagate downstream. Wake edges represented in yellow—normalized streamwise velocity of 0.75—mark the inside velocity deficit and the outer undisturbed flow field

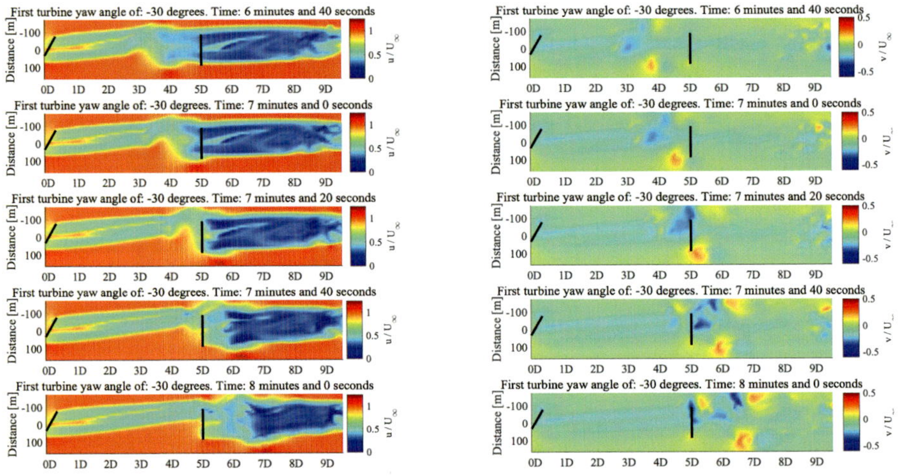

Fig. 5.6 Wake deflection dynamics (normalized streamwise and spanwise velocities, on the left and right sides, respectively): (1), (2) and (3) wake continues to propagate downstream, reaching the second turbine's rotor 2 minutes and 30 seconds later and (4) and (5) deflected wake is fully developed, and velocity deficit affecting downstream turbine is noticeably reduced

Figures 5.5 and 5.6[4] provide a quantitative overview on the wake's characteristics under the control strategy being analysed. The snapshots are taken at hub height,[5] and both the streamwise u and spanwise v velocities, normalized by the upstream velocity U_∞, are represented. These two components play a major role in the dynamics of the wake. For background information regarding details of the simulation, the reader is referred to an earlier section of this book, namely Sect. 4.1.1.

In the last graph of Fig. 5.2, the effectiveness of wake redirection control by yaw misalignment can be observed. Using greedy control (γ_{t1} set to zero) turbine 1 captures 5.49 MW and turbine 2 0.77 MW. Using an offset of $-10°$ leads to turbine 1 capturing 5.35 MW and turbine 2 0.96 MW—a total collective power increase of 0.8% from 6.26 MW to 6.31 MW. By further yawing the upstream turbine $-20°$, the last graph puts in evidence the rationale of the WRC strategy. Turbine 1 is sacrificed, generating 4.20 MW in the fully yawed setting. Turbine 2, on the other hand, compensates for the lost 1.29 MW by capturing 3.36 MW, an additional 2.6 MW with relation to the greedy control scenario. This corresponds to a power production increase of 20.77% from 6.26 MW to 7.56 MW at the wind farm level.

5.2 Data Pre-processing

Prior to computing the ROM, the input output data is pre-processed to remove the signals from their main trends, so that the models can focus on the true dynamics of the system. For the case of wake redirection control, the initial yaw misalignment of $-10°$ is removed from the input signal, and the generator power at that setting (yaw angle of $-10°$) for the upstream and downstream turbines is similarly removed from the generator power signals. The latter values, for turbines 1 and 2, are 5.49 MW and 0.77 MW, respectively (Fig. 5.7).

Up until this point, only the input-output signals were pre-processed. The pre-processing involved a resampling and removal of the mean value. Other techniques could be applied here, such as scaling of the input-output data by a certain variable, such as the variance of the signal. Ideally, the model should be agnostic to the type of scaling, and performing this step on the input-output signals may improve the results of the system identification step. For the sake of completeness, it should

[4] These snapshots were obtained using the function `hubheightcut_yaw.m`, also available in the GitHub webpage.

[5] Distance from the turbine platform to the rotor of an installed wind turbine, indicating how high the turbine stands above the ground (not including the length of the turbine blades).

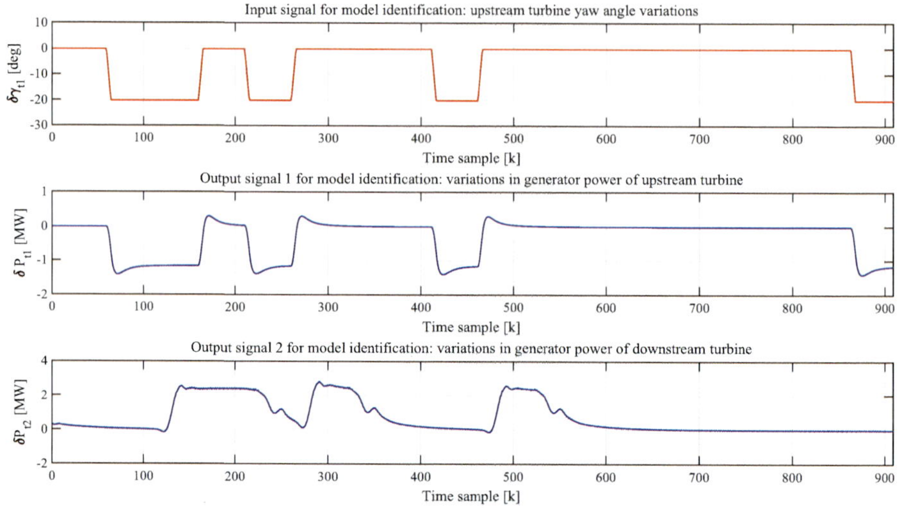

Fig. 5.7 Input-output measurements to be used for model identification for wake redirection control: detrended and resampled turbine 1 yaw angle (orange) and generator power of both turbines during identification simulation (blue). Original content in [5], used with permission of the American Automatic Control Council (AACC)

be clarified that scaling of signals was also performed initially, although not much added value was found. Given that one of the final purposes of the ROM was to design a Model Predictive Controller, it was decided to maintain complexity as low as possible to make its design straightforward. Hence, scaling of the input-output data is not performed in these examples.

In addition to pre-processing the input-output signals for the system identification step, the flow field data was also pre-processed. In the initial simulation the complete flow field was retrieved, including locations downstream of the second turbine. This is what has already been represented in Figs. 6.5 and 6.6. However, this flow field data is resampled, and only the wake between the two turbines is retrieved to build the snapshot data matrix. The rationale underpinning this choice is that the most interesting dynamics to capture are mainly concentrated between the first and second turbines.

5.3 Model Calculation

To calculate the IODMD models, the function in Listing 5.1 is used. Three input arguments are needed. The states correspond to the three dimensional flow field information concatenated in a matrix format, where each column corresponds to data for one time instant, following the snapshot methodology. For simplicity purposes, in a first attempt at using the algorithm, only the streamwise velocity component u is used to build the state matrix. Note that other velocity components can be used and the states information can be pre-processed beforehand. The input and output signals are similarly concatenated in a matrix format and serve as the remaining two input arguments. The three input matrices are then used to build their corresponding snapshot and time-shifted versions. This is depicted from lines 8–16 in Listing 5.1.

Listing 5.1 Model calculation and validation using IODMD

```
1  function [ROM]=iodmd(states, Inputs, Outputs)
2
3      %% INPUT ARGUMENTS
4      %states:   states to build DMD matrices
5      %inputs:   [U1 U2 U3 ... Un];
6      %outputs: [Y1 Y2 Y3 .... Yn];
7
8      %define necessary matrices for IODMD
9      X         = states(:,1:end-1);          %snapshot matrix
10     X_p       = states(:,2:end);            %time-shifted snapshot matrix
11
12     out       = Outputs(:,1:end-1);         %output data matrix
13     out_p     = Outputs(:,2:end);           %ime shifted output data matrix
14
15     inp       =Inputs(:,1:end-1);           %input data matrix
16     inp_p     =Inputs(:,2:end);             %time shifted input data matrix
17
18     % truncation value
19     r=100
20
21     %perform SVD of X
22     [U,S,V]=svds(X,r);
23
24     %pre-allocate matrix with fitness values for each model
25     FITje=zeros(2, r);
26
27     %pre-allocate cell with state space trajectory of each model
28     x=cell(r,1);
29
30     %loop over each singular value
31     for si=1:1:r
32
33         %take U,S,V for each number os singular values used
34         Util=U(:,1:si);
35         Sigtil=S(1:si,1:si);
36         Vtil=V(:,1:si);
37
38         %use IODMD solution
39         mat=[Util'*X_p;out]*pinv([Sigtil*Vtil';inp]);
40
41         %fetch A,B,C and D matrices
42         A{si}=mat(1:si,1:si);
43         B{si}=mat(1:si,si+1:end);
44         C{si}=mat(si+1:end, 1:si);
45         D{si}=mat(si+1:end, si+1:end);
46
47         %store state space model of size si
48         sys_red{si}=ss(A{si},B{si},C{si},D{si},2);
49
50         %evaluate fitness of model
51         [FITje,x]=evaluatemodel(sys_red,si,Inputs,Outputs,FITje,x,states,U);
52
53     end
54 end
```

A truncation value of 100 is chosen, only as a first example. This value is chosen to be large enough so as to encapsulate enough modes to recover the dominant dynamics. The purpose of this is not to compute one ROM with 100 dynamic modes, but to compute 100 ROMs, each one with an increasing mode (the last one indeed with r singular values and dynamic modes). This is achieved by performing the SVD and only retrieving the relevant information for each additional singular value to compute the ROM. The for-loop initiated in line 31 achieves this. All ROMs are stored in a cell array, each one with its corresponding singular value index, referred to as `si`. The state space model is generated using the inbuilt MATLAB command `ss`. The fitness of the model is then evaluated by using the function `evaluatemodel`, called in line 51 and shown in Listing 5.2.

The evaluation of the model is performed by using the already introduced criteria Variance Accounted For (VAF). The VAFs between the true outputs (the generator power of the upstream and downstream turbines) and estimated by the ROM are then quantified (in practical terms, this is done in line 13 of Listing 5.2). For this, the ROM outputs need to be retrieved. This is achieved by simulating the ROM using the inbuilt MATLAB function `lsim`, which evaluates the response of each of the ROM stored in the cell array `sys_red{si}` to the input signal (in this case, variations in the upstream turbine yaw angle). An initial condition is given for the simulation. To estimate it, the high dimensional flow field at the initial time instant, concatenated onto a one dimensional vector, is projected onto the low dimension space by pre-multiplying with the transpose of the $\mathbf{U}$ matrix which contains the left singular vectors of the snapshot data matrix.

Other possibilities for estimation of the initial state vector exist. An alternative method would be using the inbuilt MATLAB function `findstates`, which estimates the initial state vector by minimizing the deviations between the measured input and the measured output. Using the latter approach, however, led to mismatches between the high dimensional state vector reconstructions. In more concrete terms, it was seen that the wake reconstruction using the ROM differed from the wake from the simulation. This topic will be further discussed in the last chapter; however it should be noted that estimation of initial conditions can be performed adopting different methods with implications on model accuracy.

The remaining of Listing 5.2, from line 15 to line 53, automates the comparison between the outputs from the ROM being analysed and the output from the simulation, depicting the VAF between the two. The results of this function can be visualized for three of the ROMs computed, in Figs. 5.8, 5.9 and 5.10, each one with a different number of modes: 3, 40 and 86, respectively. The VAF of all possible models is further depicted in Fig. 5.11. Note that the represented VAF is computed with the outputs of the ROM simulated with identification data. This is achieved by (1) selecting a truncation value of $r = 100$ for the SVD, (2) computing ROMs of increasing size, (3) evaluating each one by computing the VAF for each output and (4) representing all information in one single plot, as in Fig. 5.11.

Listing 5.2 Model evaluation

```
function [FITje,x]=evaluatemodel(sys_red,si,Inputs, Outputs,FITje,x,states,U)

    %compute initial condition
    xo=U(1:size(states,1),1:si)'*states(:,1);

    %linear simulation of state space system
    [ysim, t, xout]=lsim(sys_red{si}, [Inputs]',[],xo);

    %save state space trajectory
    x{si}=xout;

    %compute variance accounted for
    FITje(:,si)=vaf([Outputs]',ysim);

    %visualise model performance (identification data)
    fig1=figure(1000+si);
    set(gcf,'color','w','Position', get(0, 'Screensize'),'Visible','on');

    %upstream turbine performance
    subplot(2,1,1)

    %plot data from simulation, first turbine
    plot([Outputs(1,1:end-1);]','b','LineWidth',1.6');
    hold on;

    %plot ROM output, first turbine
    plot(ysim(:,1),'Color',[0, 0.5, 0.0],'LineWidth',1.6,'LineStyle','--')
    grid on
    xlabel('Time instant [k]')
    ylabel(' Power [MW]')
    title(['Model fitness: generator power for first turbine. VAF of ',num2str(round(FITje(1,si),2)),' %. '])
    legend({'Real Generator Power','Model Output'},'Location','bestoutside','Orientation','horizontal')
    legend('boxoff')
    set(gca,'fontsize', 14)

    %downstream turbine performance
    subplot(2,1,2)

    %plot data from simulation, second turbine
    plot([Outputs(2,1:end-1)]','b','LineWidth',1.6'); %1
    hold on;

    %plot ROM output, second turbine
    plot(ysim(:,2),'Color',[0, 0.5, 0],'LineWidth',1.6,'LineStyle','--')
    grid on
    xlabel('Time instant [ ]')
    ylabel(' Power [MW]')
    title(['Model fitness: generator power for second turbine. VAF of ',num2str(round(FITje(2,si),2)),' %. '])
    legend({'Real Generator Power','Model Output'},'Location','bestoutside','Orientation','horizontal')
    legend('boxoff')
    set(gca,'fontsize', 14)

end
```

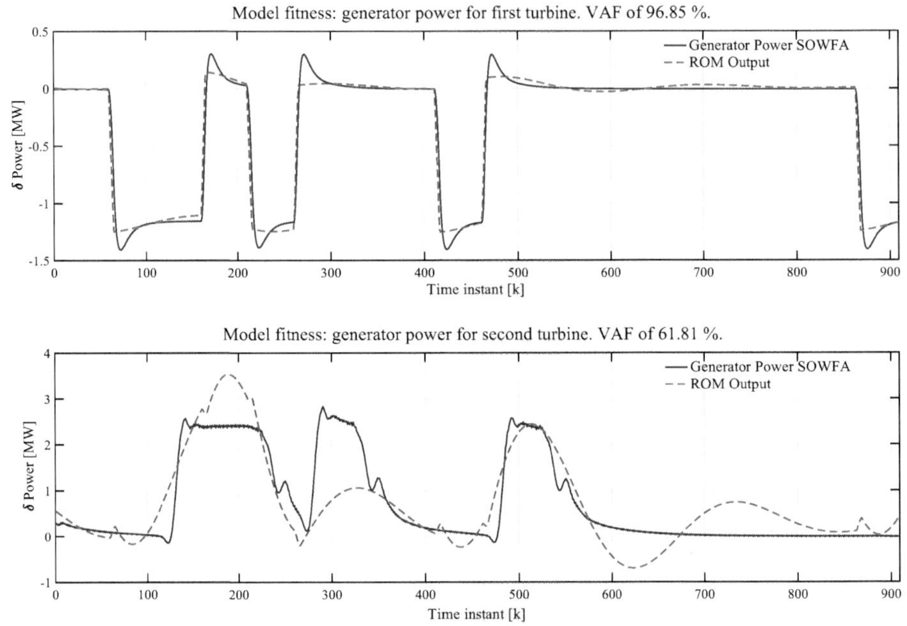

Fig. 5.8 Evaluation of Reduced Order Model performance. Model computed using three singular values and modes. The first plot shows Variance Accounted For of the first output (generator power of upstream turbine), and the second plot shows the second output (generator power of downstream turbine). Data from SOWFA is shown in blue colour. Dashed green lines depict ROM's output for a linear simulation where input identification data is used

Based on the results shown, some important remarks should be highlighted:

1. *The VAF of the first output is significantly higher than the VAF of the second output:* Regardless of the ROM shown, all of the three show a significantly higher VAF for the first output than the second, as can be seen in Figs. 5.8, 5.9 and 5.10. For the ROM with three modes, for example, the VAF of the first output is 35% points higher than the second output. This can also be seen more holistically for all ROMs in Fig. 5.11, where the VAF of the first and second outputs is depicted for each of the models. The blue line, representing the VAF of the first output, is above the red line, representing the VAF of the second output. This is because of the feed-through term in the state space model for the first output. Due to the almost instantaneous drop in power production of the upstream turbine due to yawing, a feed-through term appears in the state space ROM, i.e. the matrix $\mathbf{D}$ has a non-zero term and the input $\boldsymbol{u}_k$ modifies the $\boldsymbol{y}_k$ for the same time instant k. For the downstream turbine power, however, the input will only affect the output at a later time instant k, reducing the feed-through term to (almost) zero.
   Listing 5.3 presents the MATLAB commands which selects and shows the ROM with six singular values (for demonstration purposes). All the entries of the state space model matrices can be seen as a result of running the code snippet 5.3.

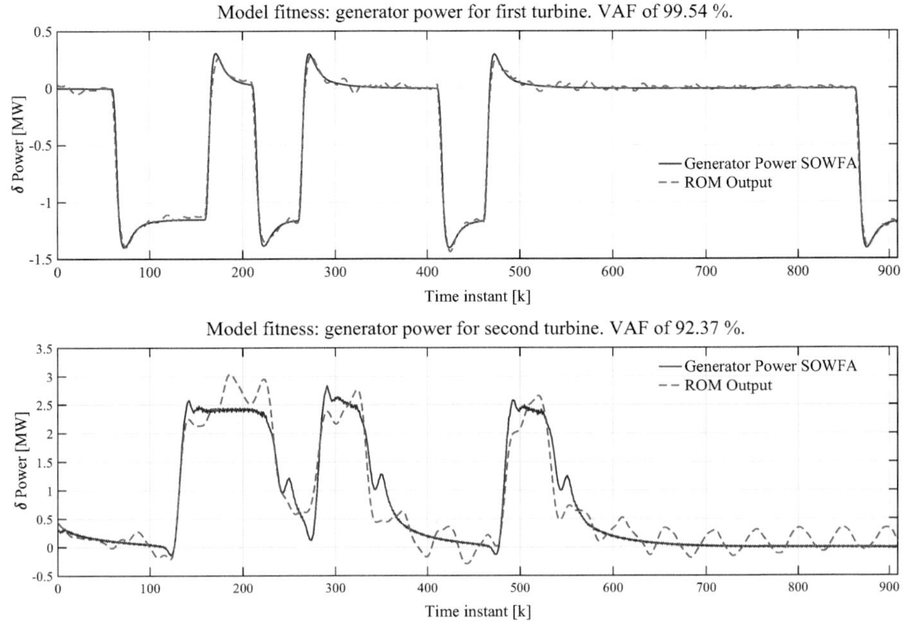

Fig. 5.9 Evaluation of Reduced Order Model performance. Model computed using 40 singular values and modes

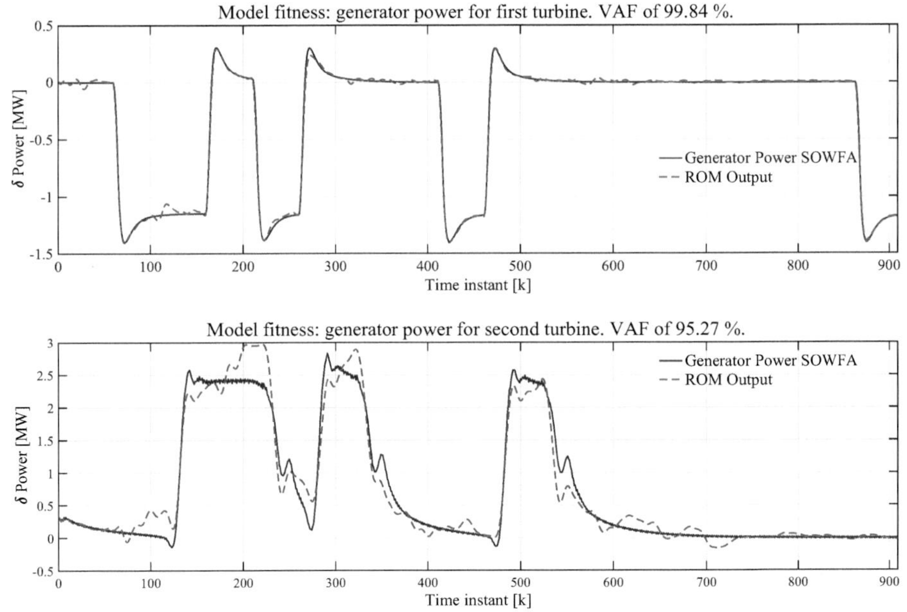

Fig. 5.10 Evaluation of Reduced Order Model performance. Model computed using 80 singular values and modes

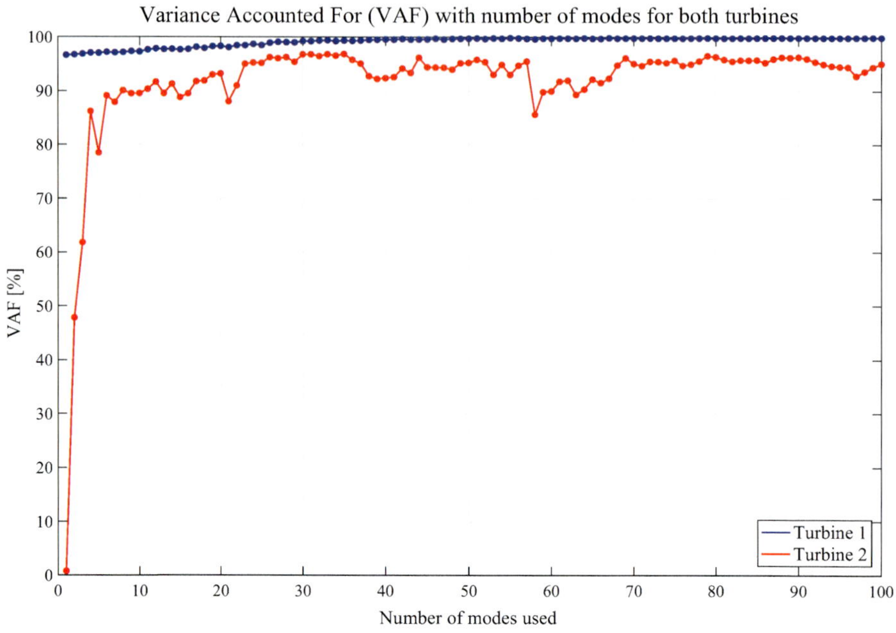

Fig. 5.11 Variance Accounted For for all ROMs of generator power of upstream turbine (blue line) and downstream turbine (red line)

The entries of matrices **C** and **D**, i.e. the mapping between states and inputs to outputs, are of special interest. It can be seen that for the first output $\boldsymbol{y}_1$, i.e. the generator power of the upstream turbine, the states barely have an influence when compared to the input. In other words, the term $\mathbf{D}_1$ is up to three orders of magnitude higher than the terms $\mathbf{C}_{1j}$, with $j = 1, 2, \ldots, 6$. For the second output $\boldsymbol{y}_2$, the term $\mathbf{D}_2$ is one order of magnitude lower than $\mathbf{D}_1$.

2. *The addition of more modes (appears) to lead to more accurate models:* It can be seen in Fig. 5.11 that as more singular values are added to build state space models of increasing size, the VAF tends to increase. This increase is more pronounced for the second output. Given the problem in hands, there is logic to this result: More modes are needed to accurately reconstruct the (non-linear) generator power dynamics of the downstream turbine, influenced by the behaviour of the wake.

Although the VAF of both outputs seems to only increase with the inclusion of more modes, this trend is most likely explained by overfitting. In more concrete terms, the addition of more modes creates a model that is able to reproduce with high accuracy the input-output relation given the training dataset it was provided but may not be able to generalize to other inputs. Therefore, validating the model becomes extremely relevant, i.e. quantitatively assessing if the already computed ROM can reproduce the behaviour of the same system for a different input-output relation. The validation of the model is discussed in the next section.

Listing 5.3 Inspection of state space matrices of reduced order model

```
%choose one of the ROMs
si=6 %choosing ROM with six modes

%show state space matrices
sys_red{si}
```

Command Window

```
>> sys_red{6}

ans =

A =
              x1          x2          x3          x4          x5          x6
   x1     0.9999  -0.0006791   -0.001612   -0.001615    0.002772   -0.009429
   x2 -0.0001547           1    -0.04503    -0.01038    0.006724    0.002879
   x3  -0.000163     0.01817      0.9953     0.03724   -0.003685     0.02872
   x4 -0.0002574    0.002033    -0.02677      0.9904    -0.07056    -0.01759
   x5  0.0001021   -0.001076    0.005819     0.04256       0.989    -0.06486
   x6 -0.0007818    0.002991    -0.02516    -0.01716     0.07799      0.9723

B =
              u1
   x1   -0.03315
   x2   -0.03638
   x3   -0.09183
   x4   -0.08759
   x5    0.04253
   x6    -0.2715

C =
              x1          x2          x3          x4          x5          x6
   y1  1.325e-05  -0.0003626   2.155e-05   -0.000323   7.596e-05  -0.0005911
   y2   0.000367   -0.007307   -0.009471    0.002153    0.001294    0.006287

D =
              u1
   y1    0.06102
   y2  -0.008362

Sample time: 2 seconds
Discrete-time state-space model.

fx >>
```

5.4 Model Validation

To validate the models a novel dataset is obtained. A new simulation in SOWFA is performed, where RBSs are again used to perturb the system within the same

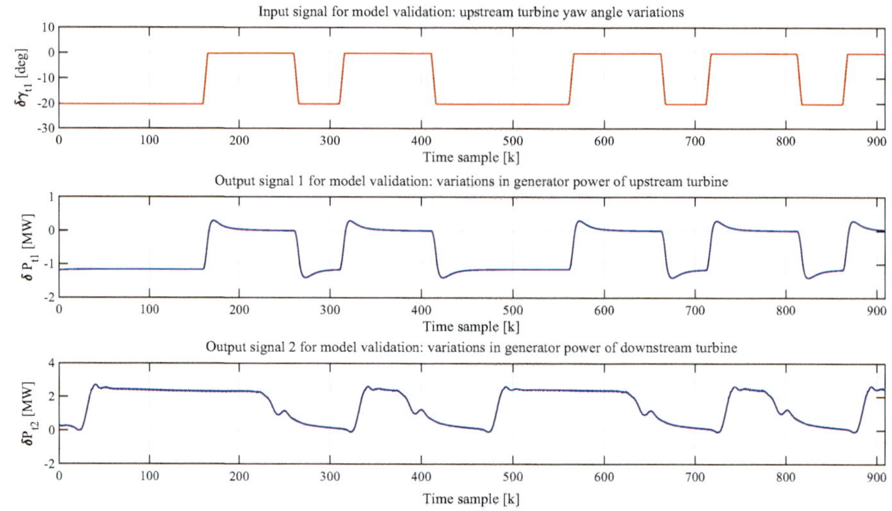

Fig. 5.12 Input-output measurements to be used for model validation for wake redirection control: detrended and resampled turbine 1 yaw angle (orange) and generator power of both turbines during validation simulation (blue). Original content in [5], used with permission of the American Automatic Control Council

range of values used for identifying the system. Input-output information is then pre-processed in a similar fashion as the data used to identify the models. Post-processed signals used to validate the models are depicted in Fig. 5.12.

Similarly to the identification case, the already computed models are simulated by using the inbuilt MATLAB function *lsim*, which simulates the discrete linear system to arbitrary inputs, in this case the upstream turbine yaw angle validation measurements. The initial condition state vector $\tilde{\boldsymbol{x}}_o$ is again computed by projecting the flow field at the initial instant onto the low dimension subspace, by doing $\tilde{\boldsymbol{x}}_o = \mathbf{U}^* \boldsymbol{x}_o$. The initial flow field from the validation dataset is used to estimate the initial conditions. The routine to validate the ROMs is shown in Listing 5.4. It is similar to the routine used for evaluating each model when using IODMD. However, now the `Inputs_val` is used to simulate the ROM, the `Outputs_val` is used to assess the VAF and the `states_valid` is passed so that the high dimensional flow field at the initial instant can be used for the initial state estimation.

Listing 5.4 Model validation

```
function [FITje_val,xstatesvalid]=validatemodels(sys_red,si,Inputs_val,
    Outputs_val,r,statesvalid,U);

%pre alocate matrix for validation fitness
FITje_val=zeros(2, length(sys_red));

%pre allocate cell for validation of state space trajectory
x_val=cell(r,1);

%loop over all state space models
for si=1:length(sys_red)
```

```
    %evaluate models for validation dataset
    [FITje_val,x_val]=evaluatemodel(sys_red,si,Inputs_val,Outputs_val,FITje_val
        ,x_val,states_valid,U);

end
```

The graph in Fig. 5.13 depicts the VAF of each one of the models computed, each one having one more state. As already observed with the identification dataset, it is clear that the dynamics of the upstream turbine are easier to model when compared to the dynamics of the downstream turbine.

Figures 5.14 and 5.15 show the performance of two ROMs simulated using the validation dataset. The first model has 37 states, whereas the second one has 59. Although the latter has more states, this does not imply that it can better recover the dynamics of interest. It is seen that the 37 states model outperforms the one with 59, with a VAF of 88% compared to 78%. It is in the engineer's best interest to balance the accuracy of the model and its complexity (this balance is often referred to as parsimony). The model with 37 states is deemed to be the most parsimonious one and better suited for the applications being discussed. This model is evaluated in greater depth by evaluating its ability to reproduce the true state space trajectory, i.e. the wake dynamics during wake redirection control, in the next section.

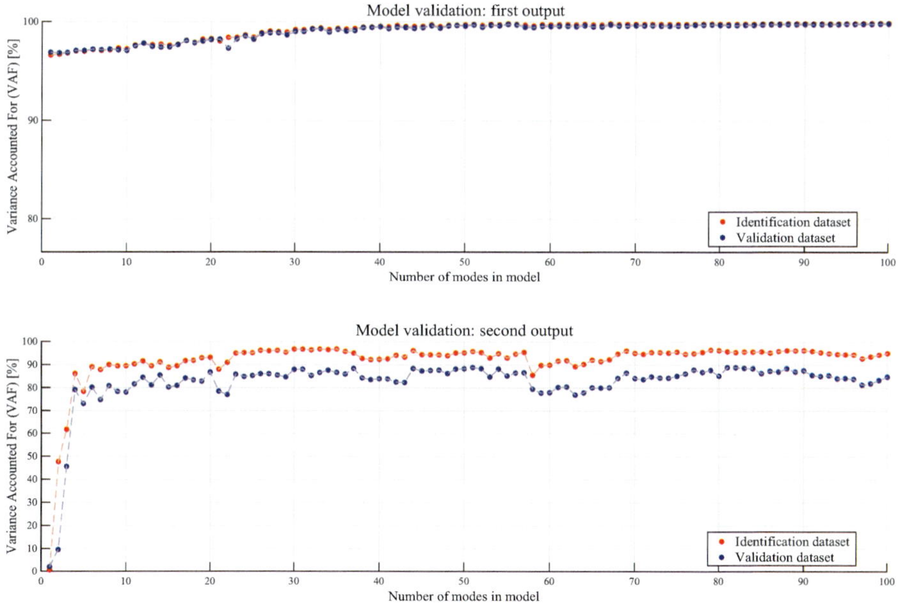

Fig. 5.13 Fitness of models. Each model, from left to right, has an increasing number of states. In red is shown the output performance using identification data and in blue using validation data. Original content in [5], used with permission of the American Automatic Control Council

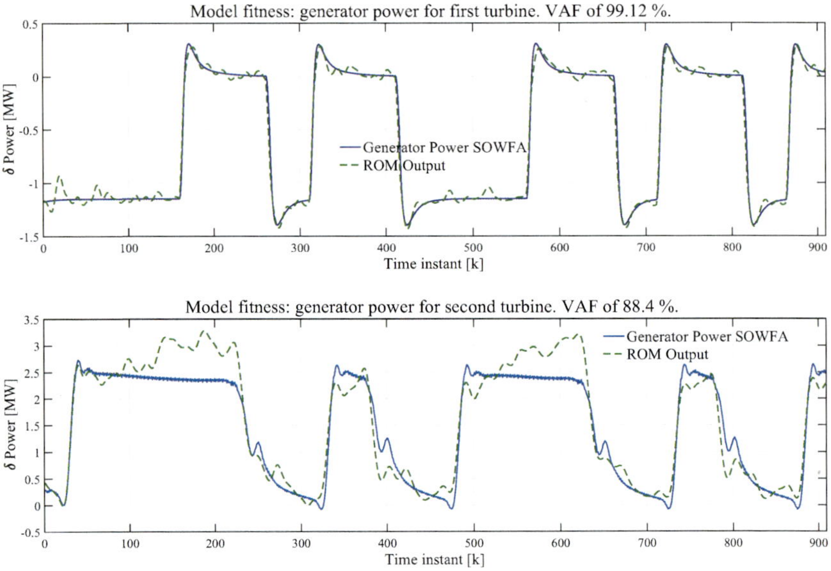

Fig. 5.14 Evaluation of Reduced Order Model performance using validation data. Model with 37 singular values validated. The first plot shows Variance Accounted For of the first output (generator power of upstream turbine), and the second plot shows the second output (generator power of downstream turbine) for the validation dataset. Data from SOWFA is shown in blue colour. Dashed green lines depict ROM's output for a linear simulation where input validation data is used. Original content in [5], used with permission of the American Automatic Control Council

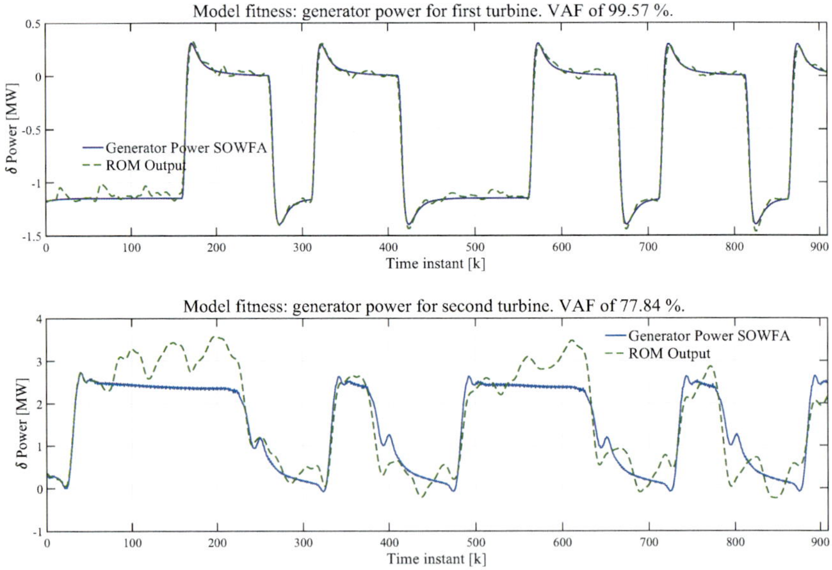

Fig. 5.15 Evaluation of Reduced Order Model performance using validation data. Model with 59 singular values validated

5.5 Wake Reconstruction

Up until this point, the focus has been on evaluating the model to predict the generated power by both turbines, the outputs chosen in the context of this analysis. In this section, the state space trajectory of the ROM is used to reconstruct the wake characteristics and its accuracy evaluated by comparing with the original information from the SOWFA simulation. For this, the ROM with 37 states is used.

Figures 5.16, 5.17, 5.18 and 5.19 put in evidence the ability of a linear state space reduced order model with 37 states to reproduce the non-linear dynamics of a wind turbine wake with tens of thousands of states by comparing the state trajectory to the real flow evolution from the simulation at hub height.

On the left side of Figs. 5.16, 5.17, 5.18 and 5.19, the velocity u is reconstructed based on the low order state trajectory[6] at key time instants when the wake is deflected. On the right, simulation data at the same time instants is shown. Several time instants following the yawing of the upstream turbine are shown to compare the evolution of the reconstruction to the actual flow field when the wake is being deflected. It can be observed, merely qualitatively in Figs. 5.16 and 5.17, that the reconstruction by the ROM is naturally limited. In Fig. 5.16 where the wake is being deflected near the upstream turbine rotor, the reconstruction and real flow seem to be in agreement. However, as the wake travels downstream and reaches the next turbine rotor in Fig. 5.17, more complex dynamics make it more challenging for a simplified linear state space model to reproduce the wake's behaviour when the latter is redirected.

Figures 5.20 and 5.21 further illustrate the model's reconstruction capabilities—for validation data—at different time intervals when the altered wake hits the downstream rotor—visualized at the downstream turbine rotor plane. Deviations are higher for the instants when the wake at the downstream rotor suffers increased alterations, as portrayed in Fig. 5.20. As the wake fully develops, deviations decrease, as illustrated in the final snapshots in Fig. 5.21.

Listing 5.5 provides a short example of how to reconstruct the wake based on the ROM state space trajectory. First, in line 3, of all the 100 models used, the best performing one, `modeltouse`, is selected. In this case, it corresponds to the model that has the highest VAF for the second output. In lines 10 and 11 the wake is reconstructed for the identification and validation datasets. The left singular vectors `U` corresponding to the singular values used for the ROM computation are

---

[6] Interpreting the state an any time instant as a point in an n-dimensional *state space*, the dynamic state response $\boldsymbol{x}(t)$ can be interpreted as a path or trajectory traced out in state space.

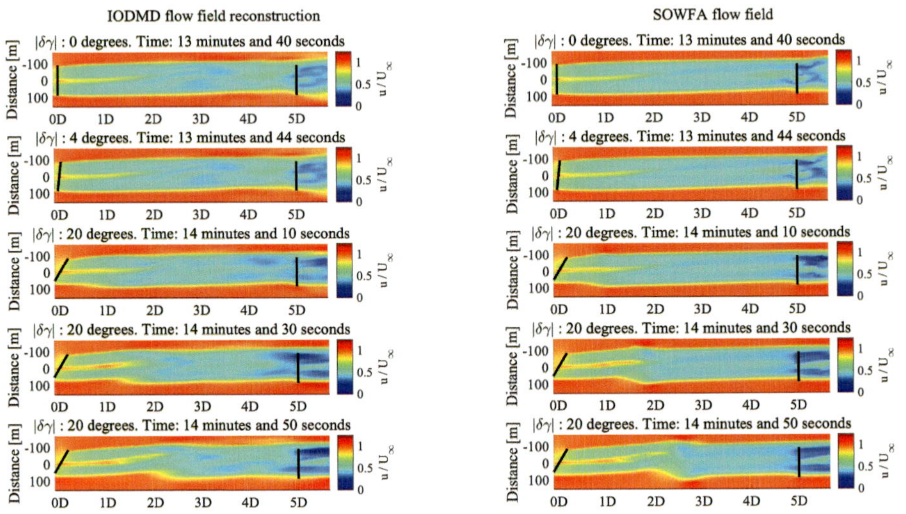

Fig. 5.16 Comparison of flow field reconstruction from low order state trajectory (left) and data from simulation (right). State trajectory obtained using validation data. Key snapshots in time are taken to illustrate the ROM's ability in predicting wake dynamics

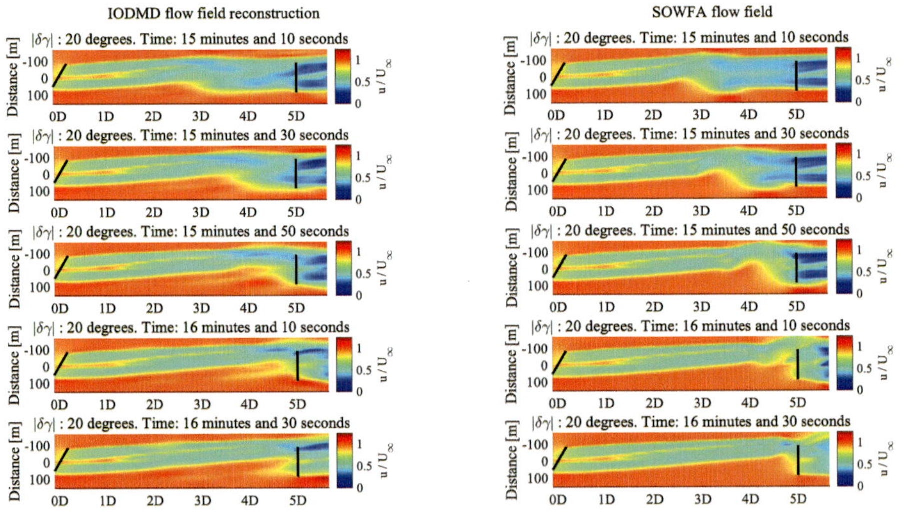

Fig. 5.17 Comparison of flow field reconstruction from low order state trajectory (left) and data from simulation (right). State trajectory obtained using validation data. Key snapshots in time are taken to illustrate the ROM's ability in predicting wake dynamics. Original content in [5], used with permission of the American Automatic Control Council

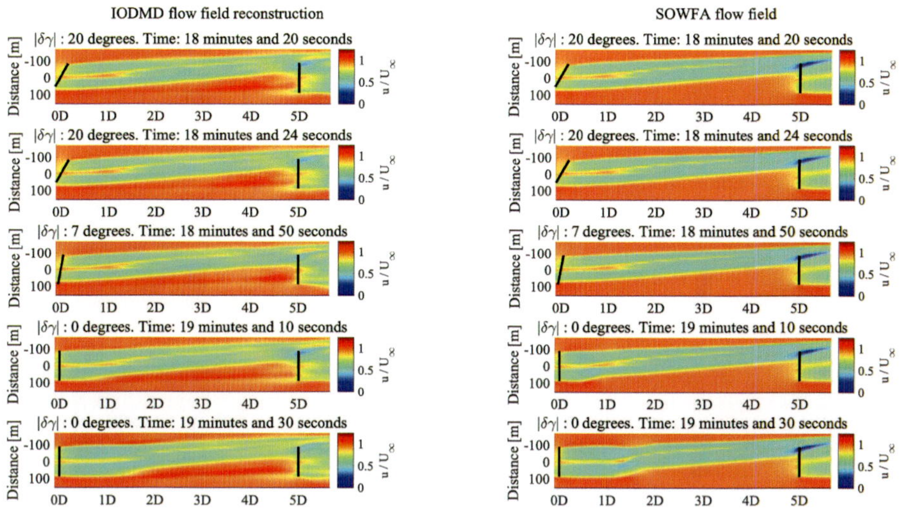

Fig. 5.18 Comparison of flow field reconstruction from low order state trajectory (left) and data from simulation (right). State trajectory obtained using validation data. Key snapshots in time are taken to illustrate the ROM's ability in predicting wake dynamics

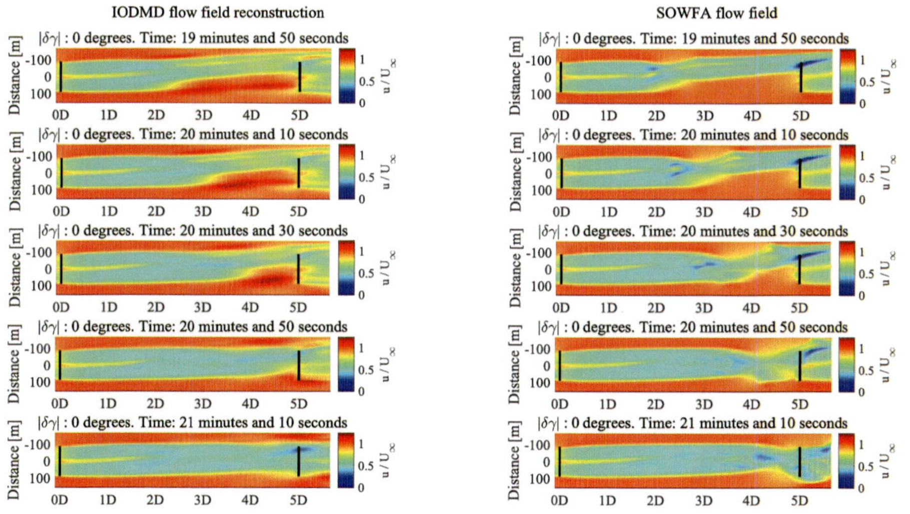

Fig. 5.19 Comparison of flow field reconstruction from low order state trajectory (left) and data from simulation (right). State trajectory obtained using validation data. Key snapshots in time are taken to illustrate the ROM's ability in predicting wake dynamics

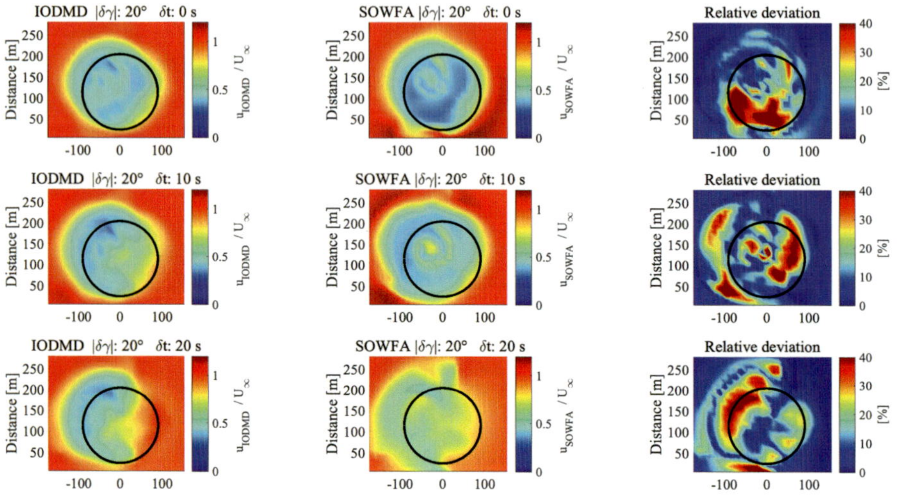

Fig. 5.20 Comparison of flow field reconstruction using reduced linear model (left) and data from SOWFA (middle) at downstream turbine rotor plane at different instances (during a total of 16 seconds). IODMD_u flow field deviation with relation to SOWFA data is shown on the right. Original content partly in [5], used with permission of the American Automatic Control Council

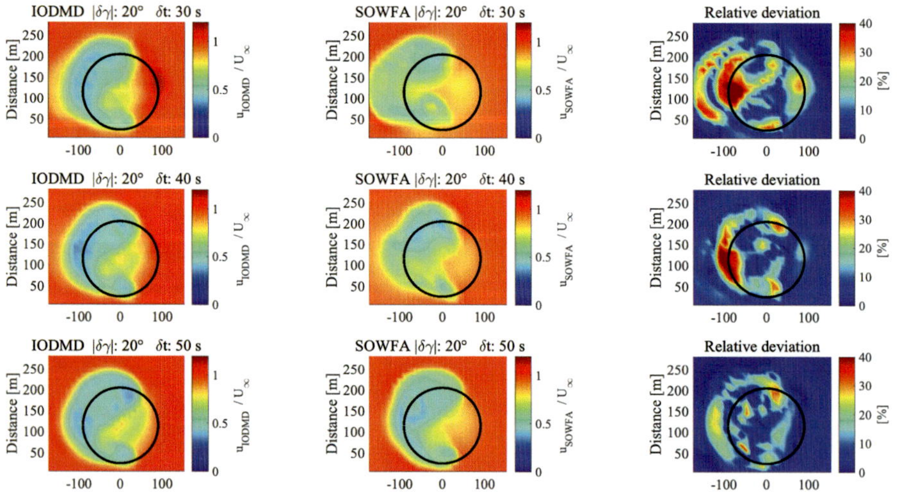

Fig. 5.21 Comparison of flow field reconstruction using reduced linear model (left) and data from SOWFA (middle) at downstream turbine rotor plane at different instances (additional 16 seconds). IODMD_u flow field deviation with relation to SOWFA data is shown on the right. Original content partly in [5], used with permission of the American Automatic Control Council

reused to project the low order state space trajectory saved in the cells `xstates` and `xstatesvalid` (for the simulations with the identification and validation datasets, respectively) to the high order state space trajectory, where the wake evolution can be visualized.

Listing 5.5 Reconstruction of flow field

```
%choose best performing model
[maxval,modeltouse]=max(FITje_val(2,1:100));

%specify size of states
n=size(states,1);

%project state space trajectory written in the low dimensional coordinates to
    higher dimensional space, where there is physical meaning
statesrebuild=U(1:n,1:modeltouse)*xstates{modeltouse}';
statesrebuildvalid=U(1:n,1:modeltouse)*xstatesvalid{modeltouse}';
```

It has been seen in the several snapshots that there are differences in the wake behaviour and the reconstruction given by the model. These differences can be quantified at each time instant by using the NRMSE error criteria, as introduced in (4.9). Figure 5.22 shows how the deviations (relative, with relation to the flow field from the simulation in SOWFA) vary throughout time for both the identification and validation datasets. Taken together, for every time instant of the simulation, such differences represent an averaged NRMSE of 4%, given that validation data is used.

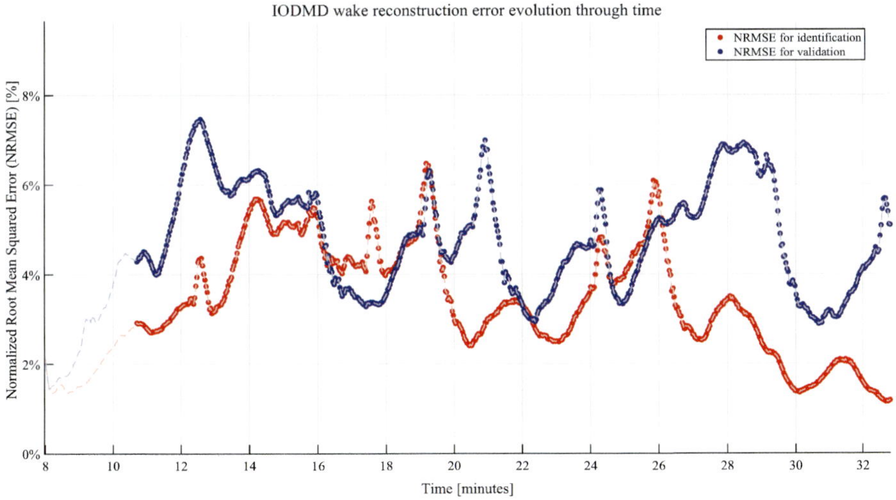

Fig. 5.22 Normalized Root Mean Squared Error (NRMSE) between the IODMD model state space trajectory reconstruction (i.e. wake reconstruction) in the high dimensional space and the wake from SOWFA for the identification (red) and validation (blue) datasets

Table 5.1 Model validation results using IODMD$_u$ algorithm

Model properties	Model obtained by IODMD				
Observable	VAF(WT2)$_{\text{max}}$	Model size	VAF(WT1)$\vert_r$	$\overline{\text{NRMSE}}$ (id)	$\overline{\text{NRMSE}}$ (val)
u	88.40%	37	99.12%	2.59%	4.00%

Listing 5.6 implements this comparison. Lines 29 through 59 compute the NRMSE for both datasets at each time instant, and the remaining of the code plots the information which is later visible in Fig. 5.22.

Using the streamwise velocity u, the IODMD$_u$ algorithm yields the results summarized in Table 5.1.

Table 5.1 has six columns: The first corresponds to the measurement used to build the data matrices and therefore the high order state $\boldsymbol{x}$. The second is the highest VAF for the second turbine generator power (the second output of the system) across all models. The dynamics of the power produced at the second turbine are more difficult to model when compared to the first turbine; therefore this measure serves to decide the model to choose as the most accurate. The third column has the size of the model—the number of states. The fourth is the VAF of the first turbine power given the chosen model. The fifth and sixth columns are the average NRMSE throughout the linear simulation. The fifth represents the ability of the model to interpolate data (given the identification dataset) and the sixth the ability of the model to predict the wake behaviour (validation data).

Listing 5.6 Deviation between IODMD model wake reconstruction and true wake throughout simulation

```
1  function [nrmse, nrmsevalid]=evaluatetimevaryingerror(X,statesrebuild,Xvalid,
       statesrebuildvalid, dirdmd,filename)
2
3  %Def: computes the NRMSE between the reconstructed wake by the IODMD model
4  %and the data form the simulation, for both the identification and the
5  %validation dataset
6
7  %Input arguments:
8  %X: the snapshot data matrix with flow field info from identification
9
10 %statesrebuild: the state space trajectory in the high space
11 %coordinates from simulation IODMD model with inputs from
12 %identification dataset
13
14 %Xvalid: the snapshot data matrix with flow field info from validation
15
16 %statesrebuildvalid: the state space trajectory in the high space
17 %coordinates from simulation IODMD model with inputs from
18 %validation dataset
19
20 %dirdmd:dir to save files
21
22 %filename: name of file to be saved when automatic exporting
23
24 %Output arguments:
25 % nrmse: time series of NRMSE for identification dataset
26 % nrmsevalid: time series of NRMSE for validation dataset
27
28 %generator time vector
29 timevecc=size(X,2);
30 timevec=1:1:timevecc;
```

```
%conversion of time instant to continuous time in simulation
for l=1:length(timevec)
    k=timevec(l)+250;
    number=k/30;
    integ=floor(number);
    fract=number-integ;
    minutos=integ;
    segundos=60*fract;
    timeveccont(l)=integ;
end

%% Calculate normalized root mean squared error for the identification data
for t=1:length(timevec)
    diff=X(:,t)-statesrebuild(:,t); %take the difference at each time instant
          for each cartesian point
    diffsquared=diff.^2; %take the square of that difference at each time step
    rmse=sqrt(sum(diffsquared)/length(diff)); %take the Root Mean Squared Error
    scale=max(X(:,t))-min(X(:,t)); %take the scaling factor for normalisation
    nrmse(t)=rmse/scale*100; %take the normalised root mean squared error %
end

%% Calculate normalized root mean squared error for validation data
for t=1:length(timevec)
    diff=Xvalid(:,t)-statesrebuildvalid(:,t);
    diffsquared=diff.^2;
    rmse=sqrt(sum(diffsquared)/length(diff));
    scale=max(Xvalid(:,t))-min(Xvalid(:,t));
    nrmsevalid(t)=rmse/scale*100;
end

%% Plot
figure550=figure;
figure550.Visible='on';

%82 is index right before frst yaw on identification data. Lets not fully
%represent the NRMSE when transient may still be present

set(gcf,'color','w','Position', get(0, 'Screensize'));
%sid=scatter(timevec(82:end),meandeltainstp(82:end),'o');
sid=scatter(timevec(82:end),nrmse(82:end),'o');
hold on
sid.MarkerFaceColor = [0.8 0.1 0.1];
sid.MarkerEdgeColor = [0.8 0.1 0.1];
pid=plot(nrmse,'LineWidth',1.1','color','red');
pid.LineStyle='- -';
pid.Color=[1 0.8 0.8];

hold on
sidvalid=scatter(timevec(82:end),nrmsevalid(82:end),'o');
sidvalid.MarkerFaceColor = [0.1 0.1 0.9];
sidvalid.MarkerEdgeColor = [0.1 0.1 0.9];
pidvalid=plot(nrmsevalid,'LineWidth',1.1','color','green');
pidvalid.LineStyle='- -';
pidvalid.Color=[0.8 0.8 1];

xlabel('Time [minutes]','fontname','times','fontsize',16);
axis([-0.3667 750 0 100])
ax=gca;
set(gca,'XTick',(0-0.3667:60:max(timevec))+1)
ax.XTickLabel = {'8','10','12','14','16','18','20','22','24','26','28','30','32
      ','34'};
set(gca,'YTick',0:2:100)
ylabel('Normalized Root Mean Squared Error (NRMSE) [%] ','fontname','times','
      fontsize',18);
set(gca,'fontsize', 18)
```

```
title('IODMD wake reconstruction error evolution through time','fontname','
    times','fontsize',22,'fontweight','normal');

grid on
hold off
ytickformat('percentage')

legend([sid sidvalid],'NRMSE for identification',...
'NRMSE for validation ',...
'Location','best','Orientation','vertical')
set(gca,'fontname','times')  % Set it to times

export_fig(figure550,strcat(dirdmd,'/image',filename),'-nocrop','-m2');
```

5.6 Dynamical Analysis

The analysis has so far focused in finding the most accurate ROM according to the pre-defined criteria of VAF and NRMSE and validating it based on data from a second simulation. The model was then explored in depth in terms of predicting the output and wake dynamics. This section further investigates the dynamical properties of the ROM used so far. Such analysis can be useful when trying to break down the underlying physical phenomena which drive the output of the ROM. The possibility to evaluate these is an advantage of using DMD related techniques and the state space model representation, as opposed to other data-driven models used in common machine learning applications that lack the same degree of interpretability.

The dynamical properties of the ROM are summarized in Table 5.2. These properties are dictated by the eigenvalues of the state matrix **A**, which are referred to as poles of the system. The poles lay in the complex plane, presenting both real and imaginary parts. The poles are converted from discrete time to continuous time by using the transformation $s = \ln(z)/T_s$, where s is the continuous time pole, z is the discrete time pole and T_s is the sampling time, in seconds. The natural frequencies[7] ω_n are computed by taking the modulus of each pole in continuous time and are then converted from rad/s to Hz. The damping ratio ξ, a dimensionless ratio that translates how oscillations of the mode decay after a disturbance, is computed by $-\cos(\text{atan2}(\Im(\ln(z)), \Re(\ln(z))))$.

The location of the poles corresponding to each of the modes can be visualized in Fig. 5.23, where they have been labelled according to the numerology in Table 5.2. In discrete time, all poles lay within the unit circle, identified with a dashed line, indicating that the system is stable (no regularization is needed).

The model can also be analysed in terms of its frequency response, this is, the steady-state response of a system to a sinusoidal input, fully described in [6]. The Bode diagram provides a graphical framework to present the frequency response characteristics of a system. In Fig. 5.24, the standard representation of the logarithmic magnitude of the system, for both outputs, is represented. The logarithmic magnitude is computed by taking the logarithm of base 10 of the amplitude ratio of the output sinusoid to the input sinusoid $20\log_{10}\left|\frac{Y(jw)}{X(jw)}\right|$, where

[7] Frequency at which a system tends to oscillate in the absence of any driving or damping force.

Table 5.2 Dynamical properties of IODMD_u model. First column entries correspond to the numerology chosen for the modes, which are ordered by increasing natural frequency. Second column entries correspond to their order in terms of eigenvectors of the state matrix **A**. Third and fourth columns have information of the modes natural frequencies and the fifth column the damping ratio information

Modes		Dynamical properties		
#	DMD	ω_n [Hz]	ω_n St [-]	ξ
1	27	0	0	1
2	32	0.0021	0.0415	1
3	37	0.0022	0.0440	1
4	30–31	0.0024	0.0466	0.6250
5	33–34	0.0048	0.0943	0.3646
6	35–36	0.0056	0.1103	0.1336
7	28–29	0.0084	0.1670	0.1771
8	25–26	0.0105	0.2072	0.0103
9	23–24	0.0133	0.2635	0.0940
10	21–22	0.0174	0.3449	0.0428
11	19–20	0.0192	0.3805	0.0205
12	17–18	0.0229	0.4532	0.0319
13	15–16	0.0254	0.5033	0.0239
14	13–14	0.0300	0.5934	0.0040
15	11–12	0.0341	0.6748	0.0032
16	9–10	0.0392	0.7752	0.0062
17	7–8	0.0433	0.8560	0.0127
18	5–6	0.0475	0.9386	0.0218
19	3–4	0.0498	0.9845	0.0082
20	1–2	0.1461	2.8904	0.0054

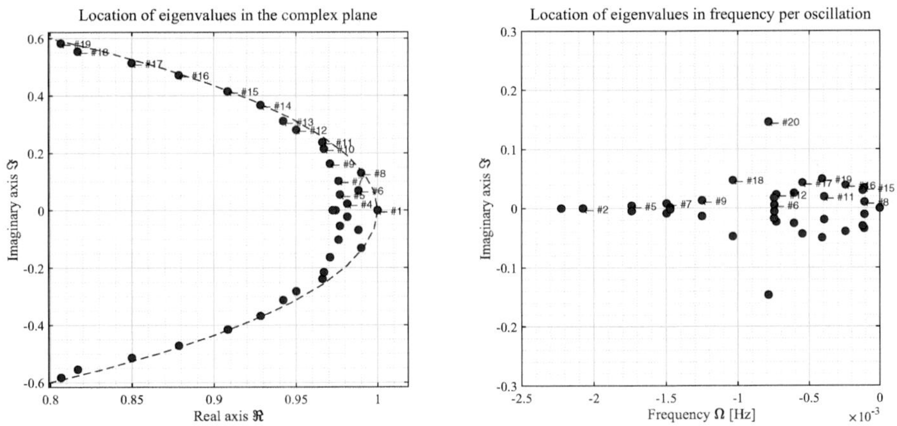

Fig. 5.23 DMD eigenvalues of the IODMD_u for wake redirection control by yaw misalignment. On the left, the discrete time eigenvalues λ on the complex plane are represented relative to the unit circle (dashed line). The area where the majority of modes are presented has been zoomed in. On the right, the eigenvalues transformed to continuous time are represented. Visualization based on the code available in [3]

ω is the frequency in rad/s. Near the frequency ω_n, a resonance peak occurs, and the damping ratio ξ determines the magnitude of this resonance peak, as can be observed by cross referencing information in Table 5.2 and Fig. 5.24. The Nyquist frequency, defined as half of the sampling rate, is marked as a vertical line.

Listing 5.7 Dynamical analysis of Reduced Order Model

```
%%Dynamical parameters of discrete time ROM

%DTU10MW rotor diameter
D=178; %m

%Upstream wind speed velocity as used in simulation
Uups=9; %m/s

%discrete time ROM time sample
dt=0.2

%eigenvalues of state space matrix A
[W,Lambda]=eig(sys_red{mn}.A);

%retrieve elements in diagonal to vector
LambdaDiag=diag(Lambda);

%conversion to continuous time
omega=log(LambdaDiag)/dt/2/pi;

%calculating adimensionalized frequency
f=abs(imag(omega))*D/Uups;

%natural frequency of continuous time pole
fnatural=sqrt(abs(imag(omega)).^2+abs(real(omega)).^2);

%damping coefficient
damping=-cos(atan2(imag(log(LambdaDiag)),real(log(LambdaDiag))));

%write eigenvectors in high order dimensional space
phi=U(:,1:best)*W;
```

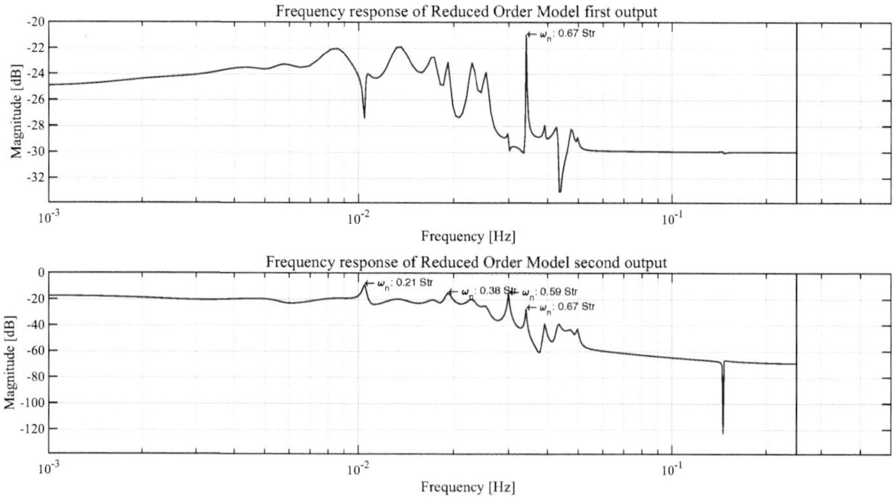

Fig. 5.24 Bode magnitude plot of IODMD$_u$ model for yaw redirection control. A number of resonance peaks have been identified, where their frequency, in Strouhal number, is depicted

Each of the eigenvalues in Fig. 5.23 corresponds to an eigenvector of the state matrix **A**, which, in turn, corresponds to a mode or coherent structure in space which inherits the dynamical properties indicated in Table 5.2. Figures 5.25, 5.26 and 5.27 allow to visualize these structures in different frequency domains. Isosurfaces of different values have been chosen so that the structure can be visualized, which are indicated in the legend. The modes, excited by certain inputs, will contribute to the overall evolution of the flow field and thus model the wake.

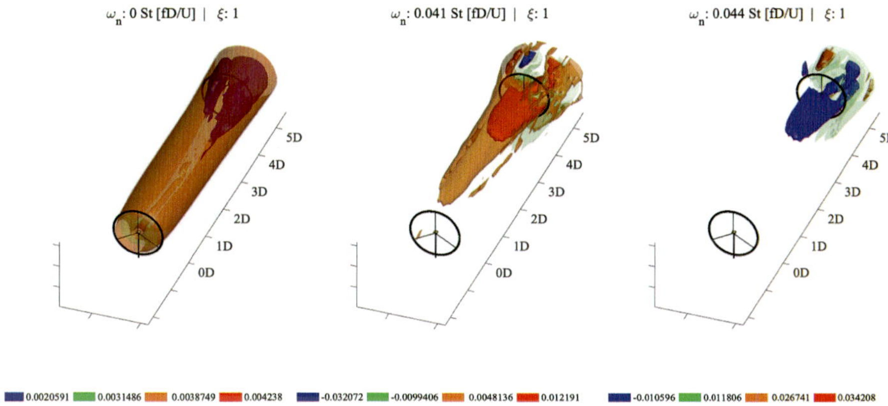

Fig. 5.25 DMD modes #1, #2 and #3. Mode #1 corresponds to a background mode that is not changing (i.e. it has zero eigenvalue). Modes #2 and #3 correspond to eigenvalues laying on the real axis

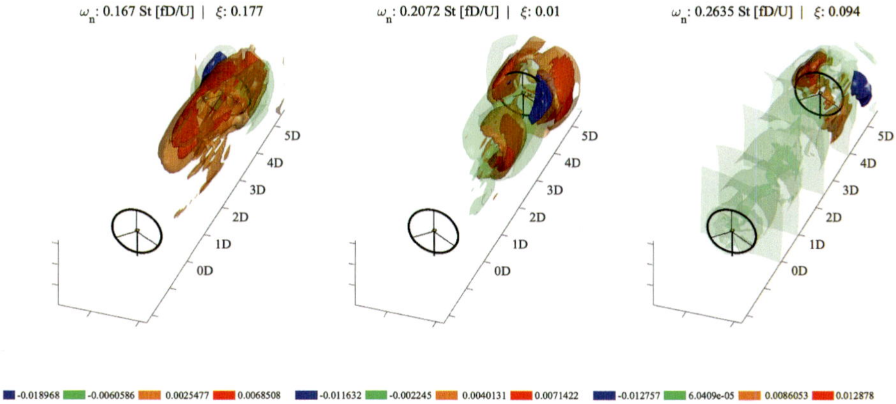

Fig. 5.26 DMD modes #7, #8 and #9, with natural frequencies comprised between 0.17 *St* and 0.26 *St*. The spatial patterns presented here appear to explain some of the dynamics more related with the downstream turbine. The structures seen here share similarities with the shapes in Figs. 5.16, 5.17, 5.18 and 5.19 as the wake is deflected near the downstream rotor

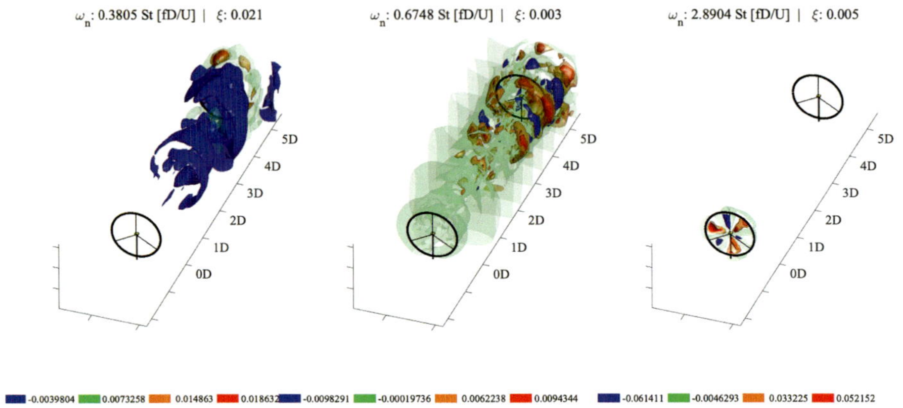

Fig. 5.27 DMD modes #11, #15 and #20, with natural frequencies comprised between 0.38 *St* and 2.89 *St*. The highest frequency mode appears to be somewhat related with fluid phenomena at the blades or rotor domain

It should be noted that these modes are merely representations of the high order eigenvectors and are not representative of the influence of the control action. Nevertheless, their study supports a deeper understanding of the wake dynamical properties, analogous to understanding the behaviour of a beam by decomposing it into different modal contributions.

5.7 Testing of Non-linear Observables

The procedure of building IODMD models where the velocity component u is used has been attempted for other control strategies (AIC), as described in [4]. One of the objectives of this work is to leverage the ideas behind the Koopman operator in order to find coordinates where the non-linear dynamics of the wake can better be described. Having the assumption in mind that other variables of interest may exist, several candidates are tested. The results are summarized in Table 5.3.

The attempt to improve the model's ability in recovering the dynamics by changing the information in the snapshot matrices appears to be fruitful. A VAF of 90% can be obtained with a linear state space model with 26 states by using the non-linear observable u^2, which represents an increase of 1.23% in VAF and a decrease in state vector size of 30% when compared to the baseline model derived with IODMD$_u$.

Table 5.3 Models validation results using IODMD algorithm with different observables. The first column has the data chosen for the states. Second, third and fourth columns have the fit criteria VAF and model size, where VAF of the second turbine generator power is the most relevant. The last two columns show the improved performance in terms of VAF and the improvements in terms of the size of model. Reduced models are less computationally expensive for the purposes of model predictive control; hence they are written in green. Table retrieved from [5], used with permission of the American Automatic Control Council

Model properties	Model obtained by IODMD			Δ IODMD$_u$	
Observable	VAF(WT2)$_{\max}$	Model size	VAF(WT1)$\|_r$	Δ VAF (WT2)$_{max}$	Δ size
v	87.12%	40	99.39%	−1.28%	+8.11%
w	87.54%	50	99.35%	−0.86%	+35.14%
u'	78.14%	38	98.08%	−10.26%	+2.70%
v'	61.30%	31	97.13%	−27.10%	−16.22%
u^2	89.63%	26	98.83%	**+1.23%**	**−29.73%**
v^2	89.58%	12	97.52%	**+1.18%**	**−67.57%**
w^2	87.85%	47	97.67%	−0.55%	+27.03%
$u \cdot v$	69.40%	15	97.58%	−19.00%	−59.46%
$u^2 + v^2$	89.42%	26	98.84%	+1.02%	−29.73%
$u^2 + v^2 + w^2$	89.36%	49	99.75%	+0.96%	+32.43%

Leveraging the ideas of the Koopman operator to increase the model's ability to reproduce the non-linear dynamics of the wind farm system has proved useful. For the case of Wake Redirection Control, replacing the initial state u by u^2 allowed to increase the VAF of the downstream turbine power signal by 1.23%. This is a small improvement but still of interest to note.

Another pattern of interest is the alike frequencies identified as relevant by IODMD, regardless of observable chosen or wind farm control strategy being studied. Comparing the frequencies of the model for the case of yaw control in Table 5.2 and the frequencies of the model for the case of collective pitch control in Table 6.2, coherent structures in the vicinity of frequencies such as 0.2, 0.59 and 0.67 *St* are visible, along with the resonance peaks in the respective frequency responses.

Other techniques such as EIODMD will be further explored in the next chapter.

This chapter exemplified the application of the data-driven techniques introduced in Chap. 3, based on the methodology described in Chap. 4, to the case of Wake Redirection Control. First, data from a high fidelity simulation is shown so that the workings of the strategy can be clearly evaluated, in terms of both input-output information and the deflection of the wake. The dataset was then processed before using it in the system identification loop. Afterwards, several models were identified following the Input-Output Dynamic Mode Decomposition approach. Each model had one additional singular value, and the influence of having models of increased size was evaluated by assessing the variations in power production of the upstream and downstream turbines resorting to the Variance Accounted For criteria. This scheme was then repeated when validating the models with a second dataset. The best performing model was then investigated. The latter was used to reconstruct the wake throughout the simulation based on the reduced order model state space trajectory projection. Differences with the simulation were assessed. In addition, the model was also investigated in relation to its dynamical properties, in an attempt to understand underlying structures responsible for the developments in the wake. Lastly, in light of the Koopman operator theory, different non-linear observables and combination of these are used to compute reduced order models, and a small benefit is seen in this approach. The same steps are taken in the next chapter, however for the case of Axial Induction Control.

References

1. N. Cassamo, J.-W. van Wingerden, *Dataset used in article: On the Potential of Reduced Order Models for Wind Farm Control: A Koop-man Dynamic Mode Decomposition Approach*. Zenodo (2020). https://doi.org/10.5281/zenodo.8287316
2. N. Cassamo, J.-W. van Wingerden, *Dataset used in article: Model Predictive Control for Wake Redirection in Wind Farms: a Koop-man Dynamic Mode Decomposition Approach*. Zenodo (2021). https://doi.org/10.5281/zenodo.8287375
3. J.N. Kutz et al., *Dynamic Mode Decomposition: Data-Driven Modelling of Complex Systems* (SIAMR, 2016)
4. J. Annoni, P.M.O. Gebraad, P. Seiler, Wind farm flow modeling using an input-output reduced-order model, in *2016 American Control Conference (ACC)* (2016), pp. 506–512
5. N. Cassamo, J.-W. van Wingerden, Model predictive control for wake redirection in wind farms: a Koopman dynamic mode decomposition approach, in *2021 American Control Conference (ACC)* (2021), pp. 1776–1782. https://doi.org/10.23919/ACC50511.2021.9482631
6. K. Ogata, *Modern Control Engineering*, 5th edn. (Prentice Hall, Upper Saddle River, 2010)

Chapter 6
Data-Driven Modelling of Axial Induction Control

In the previous chapter, the data driven methodologies introduced in Chap. 3 were applied to data from a high fidelity simulation in SOWFA where Wake Redirection Control was tested. More specifically, the Input Output Dynamic Mode Decomposition was used, where the streamwise velocity component u figured in the snapshot data matrices. A simulation was initially performed to obtain data to identify models. The wake dynamics were explored with this dataset and several models calculated. A second simulation was then performed. The dataset obtained in the latter was then used to validate the already computed models based on the pre-defined evaluation criteria. A best performing model was then chosen, and this was investigated in detail. Its ability to predict the turbine power output and the wake's behaviour was showcased. The dynamical properties of the model were then inspected. Some of the high dimensional vectors of the state space matrix **A** were visually represented in an attempt to have a better grasp of the dynamics at play. In an attempt to better capture the existing non-linear dynamics in a simplified linear model, non-linear transformations of the data used in the snapshot matrices were performed.

The methodology employed for the case of Wake Redirection Control can be easily replicated to any other dataset. This is exactly exemplified in this chapter, where the same procedure employed before is applied to data from a simulation in SOWFA where Axial Induction Control is tested. Following the systems identification loop, the starting point is to perform a first simulation where a dataset to compute models is obtained.

6.1 Simulation and Data Record

The characteristics in this second simulation are the same as the ones used in Sect. 5.1 and detailed in Sect. 4.1.1. However, in this case, the inputs used are RBS

N. Cassamo, J.-W. van Wingerden, *Data-driven Modelling of Wind Farm Flow Control Strategies*, https://doi.org/10.1007/978-3-031-84116-3_6

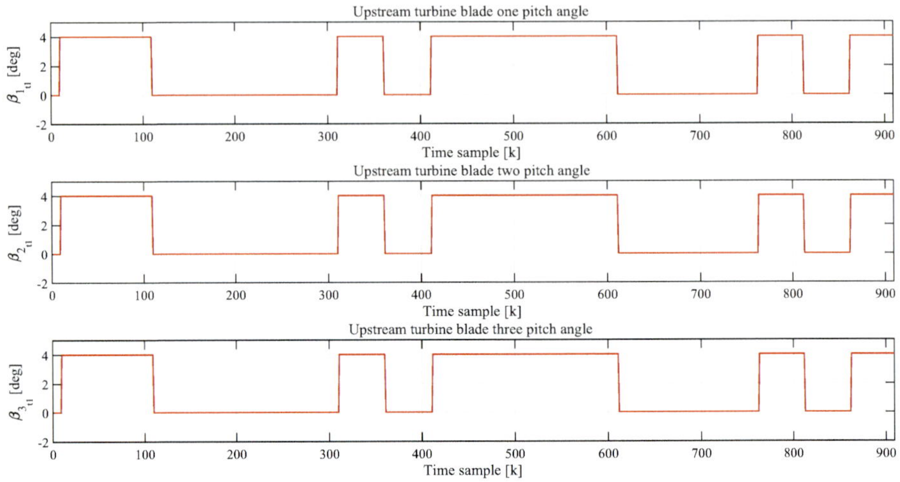

Fig. 6.1 Collective pitch angle variation of upstream turbine β_{t1} during simulation. From top to bottom, the pitch angle of each blade is represented, merely to emphasize that they are the same

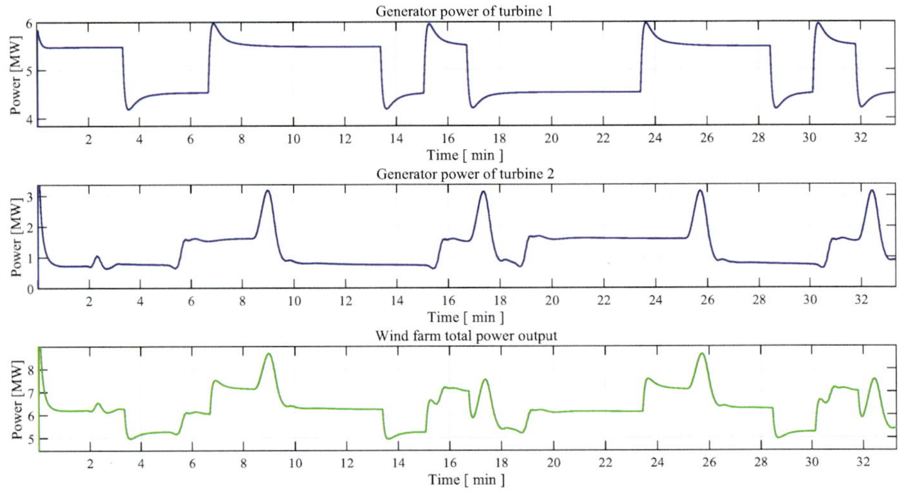

Fig. 6.2 Generator power variations of both turbines (in blue) during simulation and total generated power at wind farm level (in green)

of blade pitch angles, which are equal for all blades. The blade pitch angles vary from 0 to 4 degrees at the upstream turbine and remain unchanged (no pitching) at the downstream turbine. This is shown in Fig. 6.1.

The variations in power produced at the upstream and downstream turbine, as well as the total combined power produced, are represented in Fig. 6.2.

Power variations are explained by changes in the wake's characteristics, which can be visualized in Figs. 6.3, 6.4, 6.5 and 6.6. In these figures, the isosurface of the absolute vorticity with value 5.65 is represented.

In Fig. 6.3, the first snapshot represents the so-called greedy scenario, where the upstream wind turbine operates at the point where its production is theoretically maximum. In the snapshot immediately after, the upstream turbine is curtailed by pitching the blades 4 degrees. The power output is immediately reduced from 5.4 MW to 4.1 MW. The downstream turbine power production remains unchanged so far. In the second snapshot in Fig. 6.3, the upstream turbine blades pitch back to 0 degrees, leading to a power production increase and again changes in the wake's characteristics. This happens before the first modified wake (i.e. due to the first curtailment) reaches the downstream turbine. As soon it hits the latter, overall power production of the farm increases for a short period of time with relation to the greedy scenario. In the baseline configuration, a total power production of 6.1 MW was registered, whereas in moments where short time scale modifications take place, the power can reach values of 7.0 MW. Note that this is only verified for small time frames, as the downstream turbine benefits from more energy contained in the wake modified due to the first curtailment, whilst the first turbine pitches back its blades.

However, if such small-scale changes do not take place, increases in the order of magnitude previously seen (almost 1 MW) are not registered. For example, near minute 17 of the simulation, the upstream turbine is curtailed for a prolonged period of time. When the modified wake hits the downstream rotor, the total power production of the farm is similar to the greedy scenario.

The effectiveness of the wind farm control strategy can be observed in Fig. 6.2, which is lower when compared to yaw deflection and in accordance with evidence in literature. The power gained at the downstream turbine, 0.85 [MW], does not compensate for the power lost up the upstream turbine when the turbine blades are pitched, 0.9 [MW]. This can be seen in minutes 20–24 of the simulation. The benefits are only visible momentarily and at some instances as, for example, from minutes 7 to 9.

The velocity deficit in the wake during the simulation can also be seen in Figs. 6.5 and 6.6. On the left side of the figures, the velocity deficit in the wake is represented by visualizing the streamwise velocity downstream of the first turbine, normalized by the incoming wind speed. On the right side, the spanwise velocity is represented for completeness. From Fig. 6.5 it is evident how the velocity deficit in the wake of the first turbine is reduced as Axial Induction Control is employed. In other words, as the blades pitch 4 degrees, the velocity in the wake increases.

All the graphical material is showed here with the intent that the reader may have the most concrete and tangible understanding of the modifications in the wake's characteristics as a consequence of changing the upstream turbine control settings, based on the results from the high fidelity simulation performed.

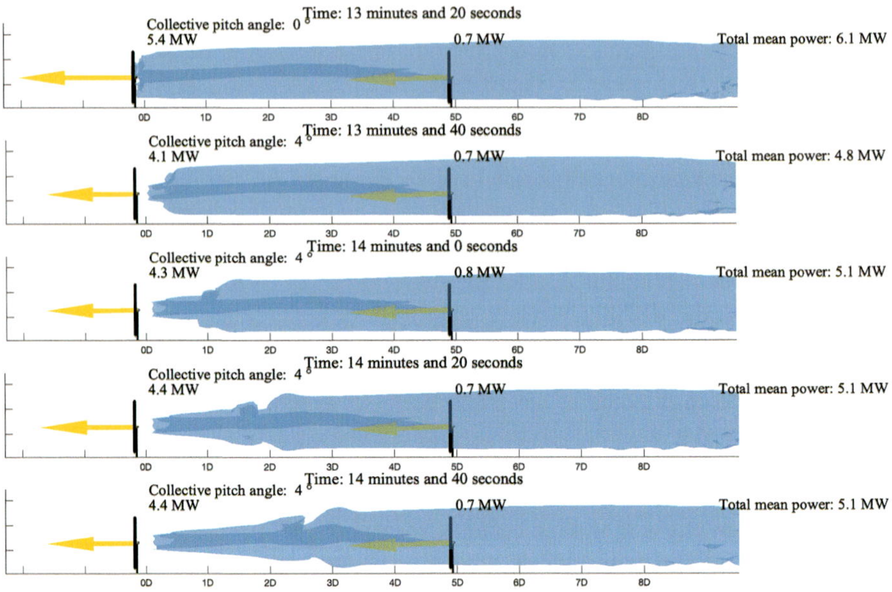

Fig. 6.3 Wake dynamics: (1) greedy control operating point, (2) blades pitch 4 degrees and power captured by upstream turbine decreases (3), (4) and (5) wakes propagates downstream

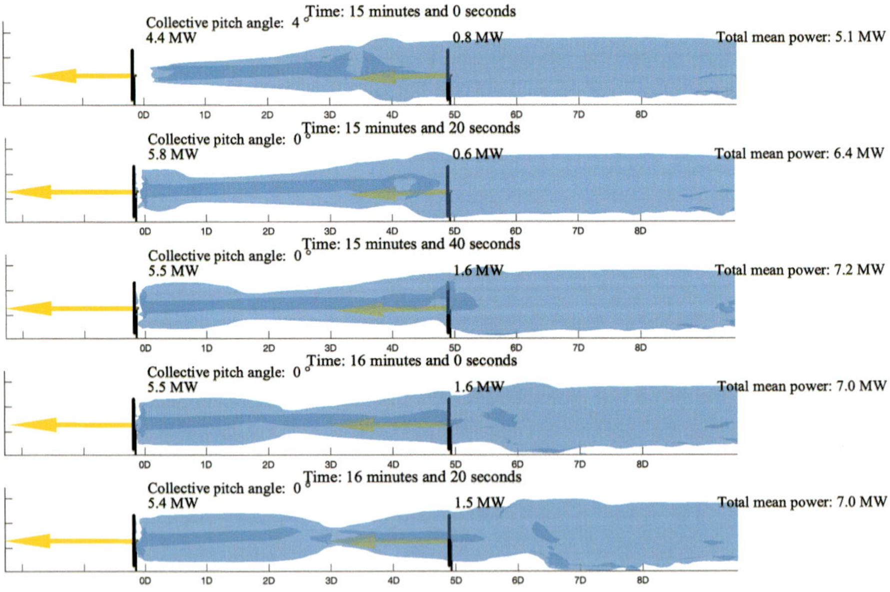

Fig. 6.4 Wake dynamics: (1) wake continues to move downstream, (2) upstream turbine blades pitch back to 0 degrees and collective captured power increases to greedy control levels, (3) altered wake hits downstream rotor and collective power increases (4) and (5) altered wake continues to propagate downstream. This latest control configuration explains the momentarily gains seen

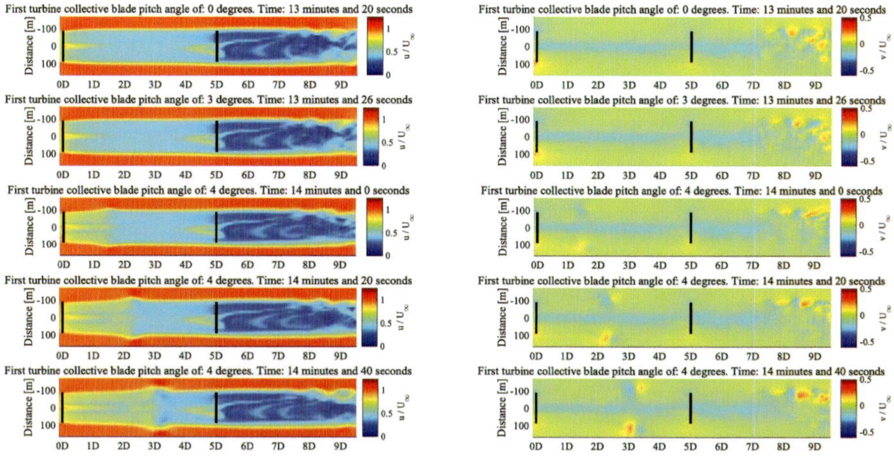

Fig. 6.5 Wake deflection dynamics (normalized streamwise and spanwise velocities, on the left and right sides, respectively): (1) and (2) first turbine starts pitching blades, (3) blades of upstream turbine are fully pitched 4 degrees and wake shows first signs of increased wind speed (4) and (5) wake continues to propagate downstream

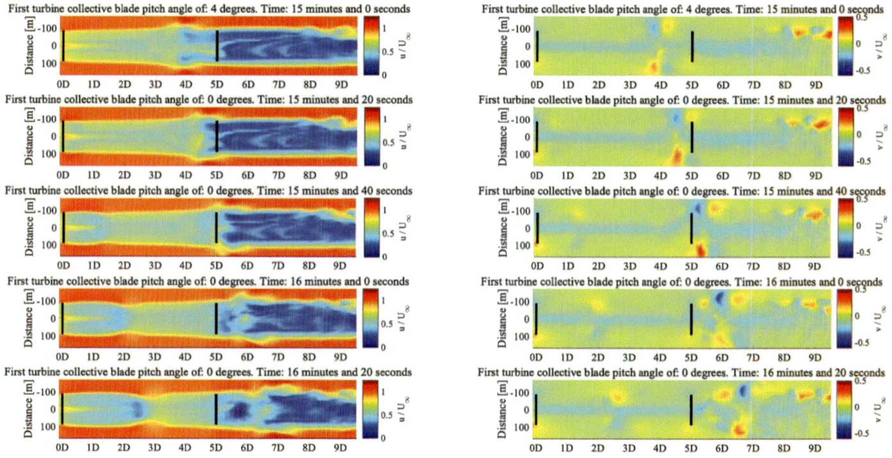

Fig. 6.6 Wake deflection dynamics (normalized streamwise and spanwise velocities, on the left and right sides, respectively): (1), (2) and (3) wake continues to propagate downstream, reaching second turbine's rotor approximately 2 minutes and 20 seconds later. Blades pitch back to 0 degrees and wake shows decreased velocity again (4) and (5) modified wake continues propagating downstream

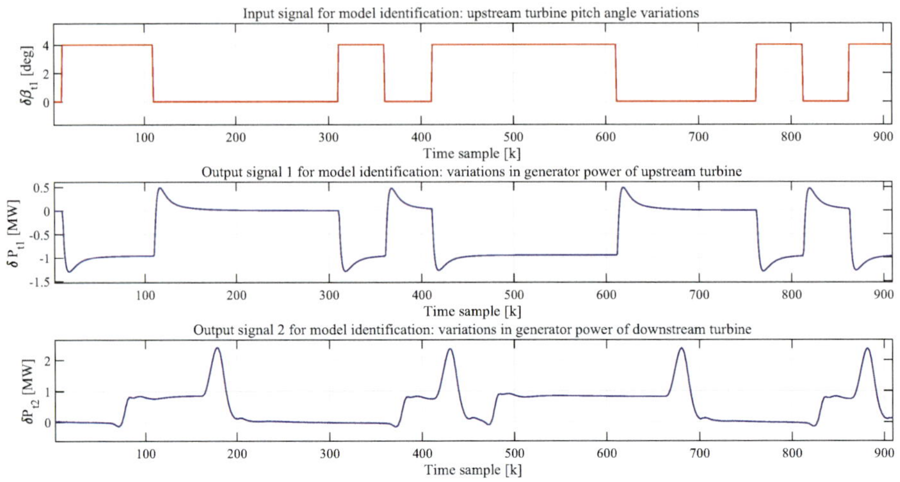

Fig. 6.7 Input-output measurements to be used for systems identification: detrended and resampled upstream turbine collective blades pitch angle and generator power of both turbines during simulation

6.2 Data Pre-processing

Data is pre-processed before calculating models. This procedure is similar to the one used for pre-processing data from the Wake Redirection Control simulation. Turbine data is resampled and the power production in steady state removed from the signals. For the case of Axial Induction Control the generator power in steady state, i.e. when no collective pitch angle is applied, corresponds to 5.48 MW for the upstream turbine and 0.77 MW for the downstream turbine.

The processed information is visible in Fig. 6.7, which is directly used, along with the states (the flow field information concatenated into the snapshot matrices), for the systems identification task. Note that no pre-processing of the flow field has been implemented.

6.3 Model Calculation

Similarly to the approach taken for the case of Wake Redirection Control, a fixed number of models are computed with an increasing number of singular values used. For each model, the VAF of both outputs is evaluated. This is represented in Fig. 6.8. Note that only the streamwise information of the flow field (the u component) was included in the snapshot matrices.

The two trends seen in the output of all models, i.e. the increase in VAF as more singular values are added and the higher VAF for the first output when compared to

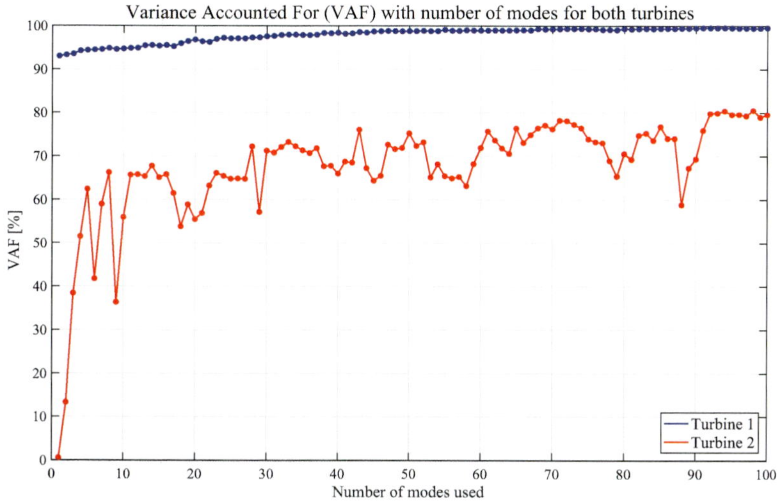

Fig. 6.8 Variance Accounted For for all ROMs of generator power of upstream turbine (blue line) and downstream turbine (red line). Identification dataset for Axial Induction Control simulation used

the second one should come as no surprise at this point. This was also the case when embedding the dynamics of Wake Redirection Control into different ROMs.

However, there is a clear difference worth noticing: The VAF of the downstream generator power for all models is lower than in the case of yaw misalignment. In addition, although it increases as more singular values are added, the rate at which it does is also lower than the previous case. A possible explanation for this difference is the fact that the non-linearity in the dynamics of the downstream generator power signal is more challenging to embed in the ROM.

This can be observed by evaluating the response of a certain ROM. Note that this response is the result of a linear simulation using as input signal the same blade collective pitch angles as the one used to calculate the ROM. The outputs of the ROM with 96 singular values are represented in Fig. 6.9 to illustrate this. This model is chosen as it is one with the highest VAF of the second output, despite the fact that most likely it has been overfitted to the data. It can be seen that the spikes in the downstream turbine power production are not that well captured (despite the first one being). This type of non-linear behaviour in the generated power in such a short time scale is not as pronounced as in the Wake Redirection Control simulation, which can explain the lower VAF found in the ROM computed from the Axial Induction Control dataset.

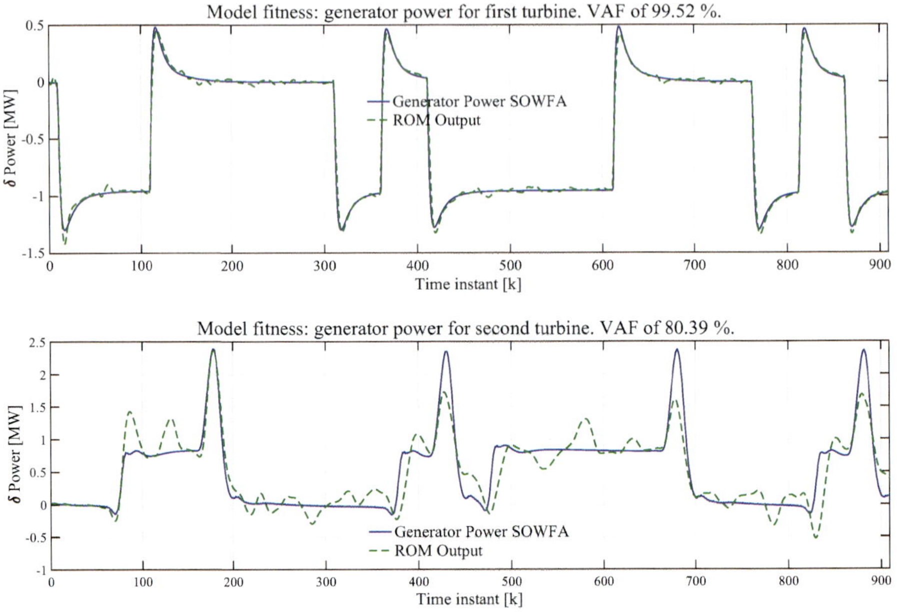

Fig. 6.9 Evaluation of Reduced Order Model performance. Model computed using 96 singular values and modes. The first plot shows Variance Accounted For of the first output (generator power of upstream turbine), and the second plot shows the second output (generator power of downstream turbine). In blue colour data from SOWFA is shown. Dashed green lines depict ROM's output for a linear simulation where input identification data is used

6.4 Model Validation

The previous computed ROMs are then validated, following the system identification loop. In accordance with the modelling approach taken for Wake Redirection Control, a second simulation is performed, where the range of the input signals, the pitch angle of the upstream turbine, is maintained the same. The input-output information retrieved from the second simulation and used to validate the models is represented in Fig. 6.10.

This first exercise of using the IODMD_u algorithm (i.e. the IODMD algorithm where the snapshot matrices contain information of the streamwise u velocity component) to compute and further validate ROMs to model Axial Induction Control yielded results of interest worth emphasizing:

1. *The best performing model has a VAF of 73.59% for the downstream turbine generated power.* The performance of the best model can be seen in Fig. 6.12. Using IODMD_u, where velocity u is chosen as state, the best model has a VAF of 74%, which is lower compared to the model used for yaw control computed in Chap. 5, which presented a VAF of 88%. It is also relevant to re-emphasize the challenge in modelling the spikes in power production at the downstream turbine.

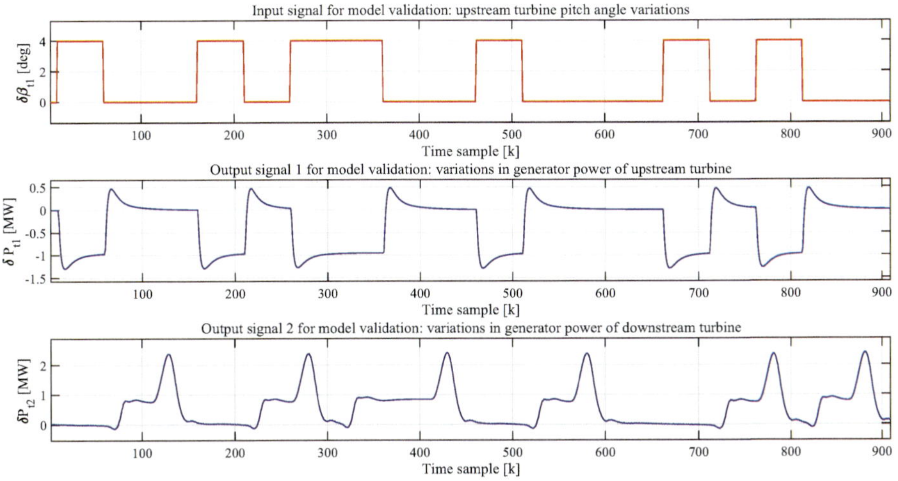

Fig. 6.10 Input-output measurements to be used for model validation: detrended and resampled upstream turbine collective blades pitch angle and generator power of both turbines during simulation.

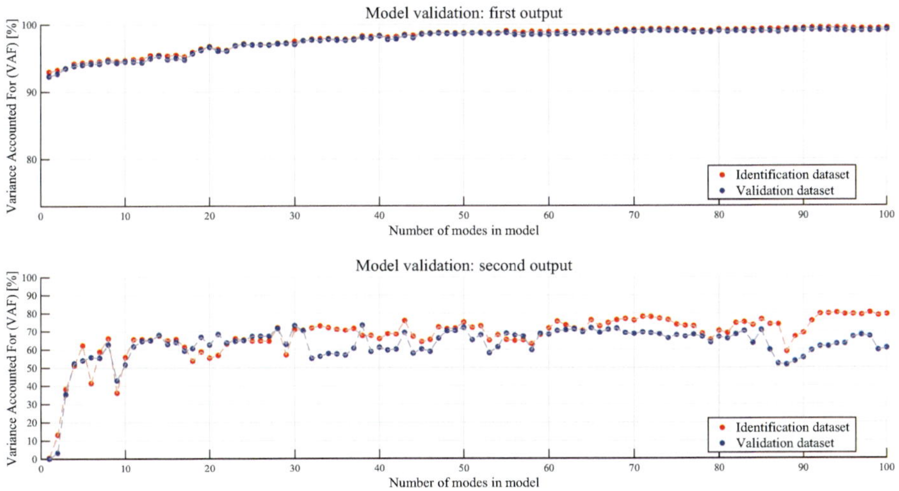

Fig. 6.11 VAF of models. Each model, from left to right, has an increasing number of states. In red is shown the output performance using identification data and in blue using validation data. Original content in [1]

2. *A certain number of modes are required to properly recover the dynamics.* In Fig. 6.11 it is shown that to reach a VAF of 70%, at least 28 modes are needed. This again indicates the increasingly difficult task of modelling the dynamics in the case of pitch control, when compared to yaw control. In the latter, only four modes are needed to reach a VAF of 80%.

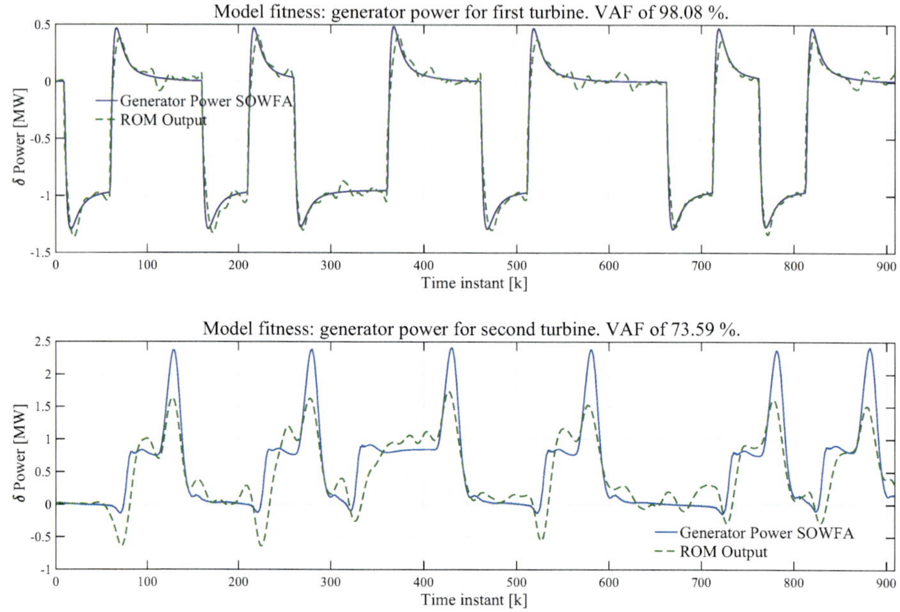

Fig. 6.12 Evaluation of Reduced Order Model performance using validation data. A model with 38 states showing the best performance as measured by the Variance Accounted For criteria. The first plot shows VAF of the first output (generator power of upstream turbine), and the second plot shows second output (generator power of downstream turbine) using validation dataset. Data from SOWFA is shown in blue colour. Dashed green lines depict ROM's output for a linear simulation with inputs from validation dataset

3. *Adding more modes does not directly lead to more accurate models.* This pattern is recognizable in Fig. 6.11, where the gap between VAF for the same models under identification and validation data widens as more modes are used (Fig. 6.12). This is possibly explained by overfitting, where the models response corresponds to the dataset used for identification. An example is models with more than 90 states. The detailed ROM response can be observed in Fig. 6.13. It can be observed in the latter that the model's response, with relation to the best performing one in Fig. 6.12, shows a more unstable behaviour, as shown by the larger dips in power production of the downstream turbine.

6.5 Wake Reconstruction

The IODMD$_u$ model can also be evaluated in its ability to reproduce the wake's characteristics. Similarly to the analysis performed for the yaw control strategy, the full velocity information at all points throughout the complete linear simulation is obtained from the reduced state vector.

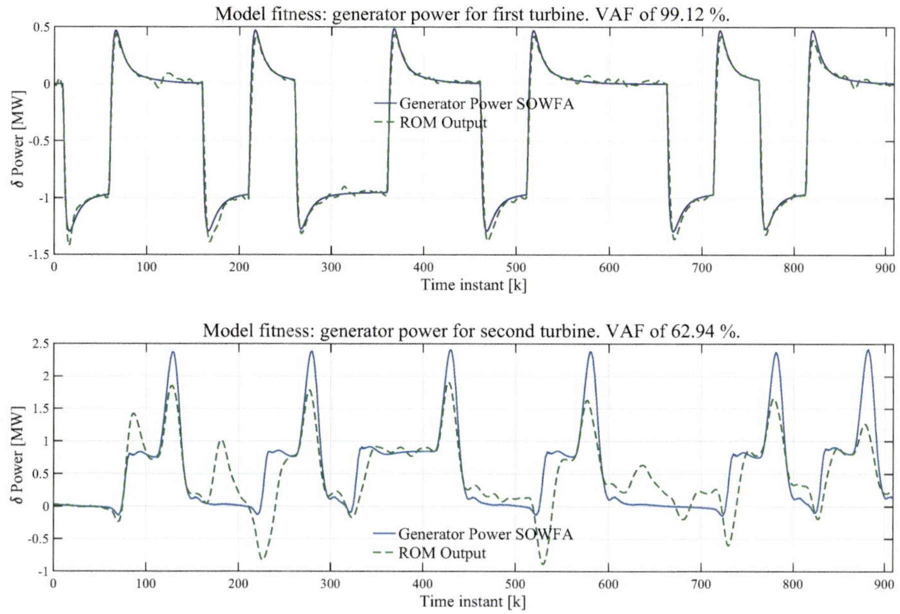

Fig. 6.13 Evaluation of Reduced Order Model performance using validation dataset. Results of linear simulation of ROM with 94 states and comparison with output from SOWFA

To illustrate the ability of the model to reproduce the wake's characteristics, the reconstruction is compared to the original flow field information retrieved in SOWFA at some key time instances. This comparison is illustrated over the next pages. In Figs. 6.14 and 6.15 the reconstructions are compared at hub height over almost 3 minutes. The time window is chosen to specifically illustrate the prediction of the flow field evolution by the ROM when the upstream turbine blades are pitched 4 degrees and then return to the 0 degrees position. Qualitatively, it can be seen that the comparison is overall accurate, in the sense that the ROM appears to encapsulate the overall wake deficit at each time instant and how it evolves over time. Nevertheless, the latter observation cannot be generalized to the entirety of the wake. For example, the outer edges of the modified wake which evolves downstream are not as sharply reconstructed as they appear in the high fidelity simulation. This region appears to be reconstructed in a much more blurred manner by the ROM. This is perhaps due to the non-linearities arising at these locations.

Figures 6.16 and 6.17 further illustrate how the IODMD reconstruction compares to SOWFA data at the downstream turbine rotor plane. An observation to note is that the gap between SOWFA information of streamwise velocity and corresponding model reconstruction widens at certain instants, namely when the modified wake hits the downstream rotor. This is especially noticeable in the last of the three snapshots in Fig. 6.16 and the first snapshot in Fig. 6.17. In the other snapshots the wake is fully developed, and IODMD$_u$ appears to better reconstruct its properties.

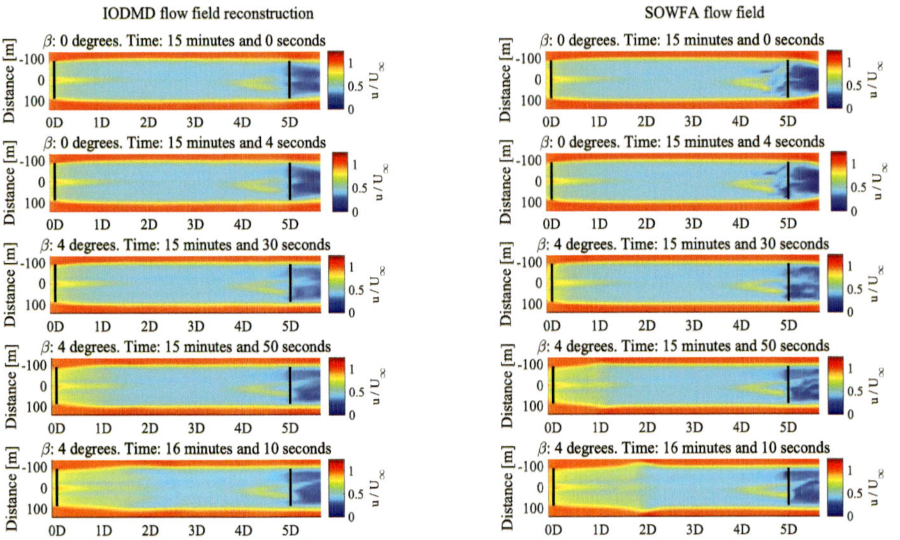

Fig. 6.14 Comparison of flow field reconstruction from low order state trajectory (left) and data from simulation (right). State trajectory obtained using validation data. Key snapshots in time are taken to illustrate the ROM's ability in predicting wake dynamics

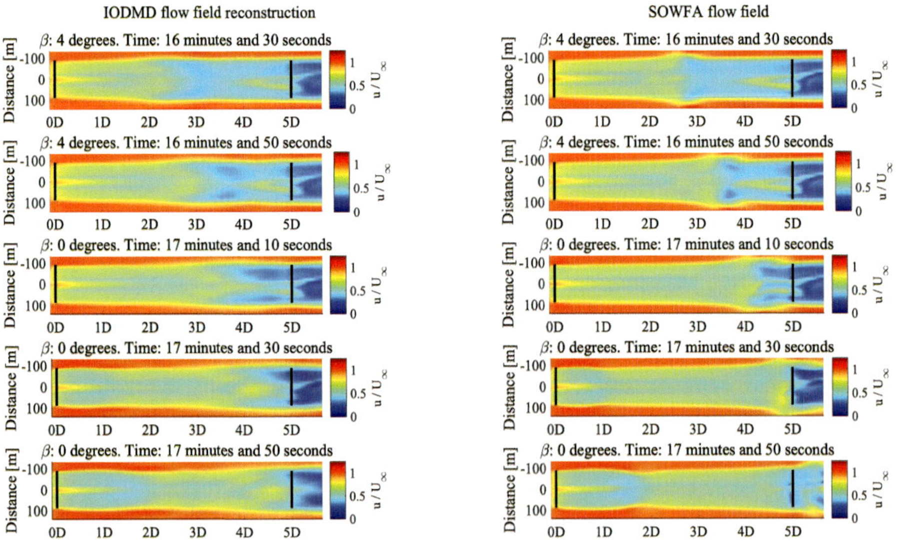

Fig. 6.15 Comparison of flow field reconstruction from low order state trajectory (left) and data from simulation (right). State trajectory obtained using validation data. Key snapshots in time are taken to illustrate the ROM's ability in predicting wake dynamics

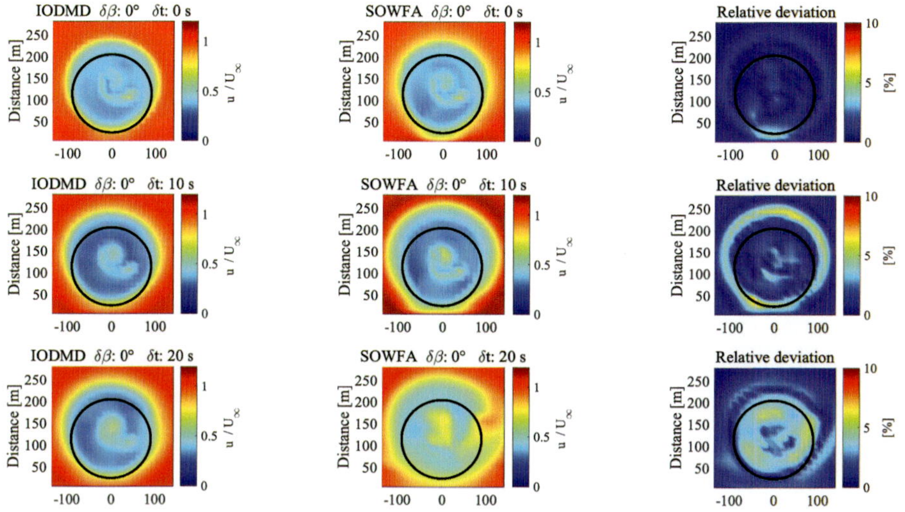

Fig. 6.16 Comparison of flow field reconstruction using reduced linear model (left) and data from SOWFA (middle) at downstream turbine rotor plane at different instances. IODMD$_u$ flow field deviation with relation to SOWFA data is shown on the right

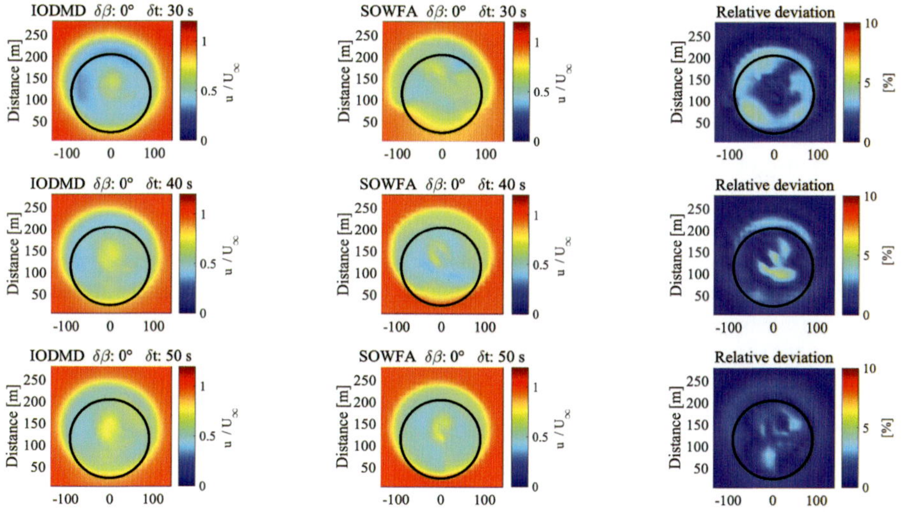

Fig. 6.17 Comparison of flow field reconstruction using reduced linear model (left) and data from SOWFA (middle) at downstream turbine rotor plane at different instances. IODMD$_u$ flow field deviation with relation to SOWFA data is shown on the right

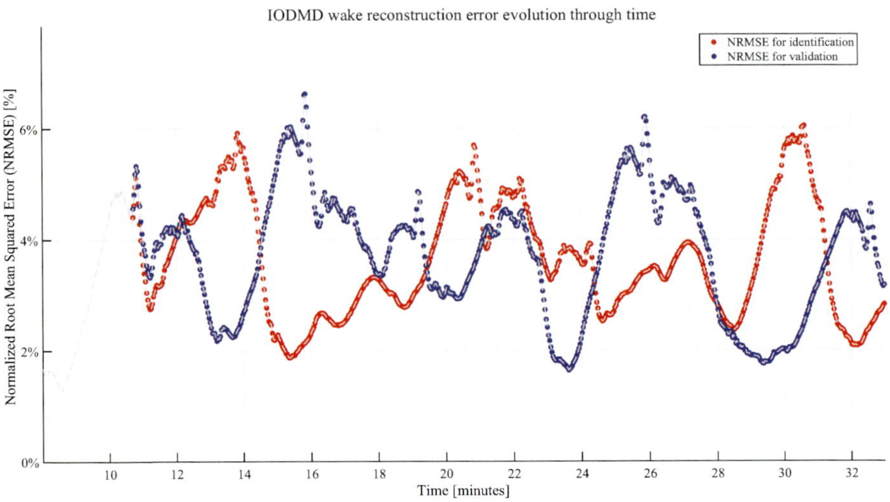

Fig. 6.18 Normalized Root Mean Squared Error (NRMSE) between the IODMD model state space trajectory reconstruction (i.e. wake reconstruction) in the high dimensional space and the wake from SOWFA for the identification (red) and validation (blue) datasets

Table 6.1 Key performance indicators of most accurate ROM using IODMD algorithm and validation dataset

Model properties	Model obtained by IODMD				
Observable	VAF(WT2)$_{max}$	Model size	VAF(WT1)$\|_r$	$\overline{\text{NRMSE}}$ (id)	$\overline{\text{NRMSE}}$ (val)
u	73.59 %	38	98.08 %	3.66%	3.75%

This can also be observed in quantitative terms in Fig. 6.18, where the criteria chosen to evaluate the discrepancies with the ROM is the NRMSE. The existing peaks in the NRMSE signal most likely coincide with modifications in the wake's characteristics, which can reach values almost as high as 8 %.

Table 6.1 summarizes the key performance metrics of the most accurate ROM, as measured by the VAF of the generated power in the downstream turbine. The average NRMSE over the complete simulation is of 3.75% for validation data, showing that the model can be used to predict the wake's characteristics.

6.6 Dynamical Analysis

The dynamical properties of the ROM are summarized in Table 6.2. These properties are taken from the corresponding poles on the complex plane, whose location dictates the dynamics of the corresponding mode. The poles' location can be seen in Fig. 6.19, in both discrete time and continuous time. The modes have been ordered by their natural frequency.

Table 6.2 Dynamical properties of IODMD$_u$ model for Axial Induction Control

Modes		Dynamical properties		
#	DMD	ω_n [Hz]	ω_n St [–]	ξ
1	27	0	0	1
2	34	0.0014	0.0283	1
3	37–38	0.0030	0.0585	0.7608
4	35–36	0.0032	0.0637	0.2041
5	32–33	0.0064	0.1268	0.3072
6	30–31	0.0088	0.1735	0.0275
7	28–29	0.0109	0.2162	0.1192
8	25–26	0.0129	0.2559	0.0497
9	23–24	0.0158	0.3124	0.0712
10	21–22	0.0194	0.3830	0.0140
11	19–20	0.0207	0.4087	0.0278
12	17–18	0.0247	0.4884	0.0358
13	15–16	0.0263	0.5204	0.0280
14	13–14	0.0301	0.5959	0.0066
15	11–12	0.0343	0.6787	0.0025
16	9–10	0.0404	0.7996	0.0026
17	7–8	0.0472	0.9331	0.0100
18	5–6	0.0560	1.1071	0.0040
19	3–4	0.1479	2.9251	0.0091
20	1–2	0.2249	4.4473	0.0482

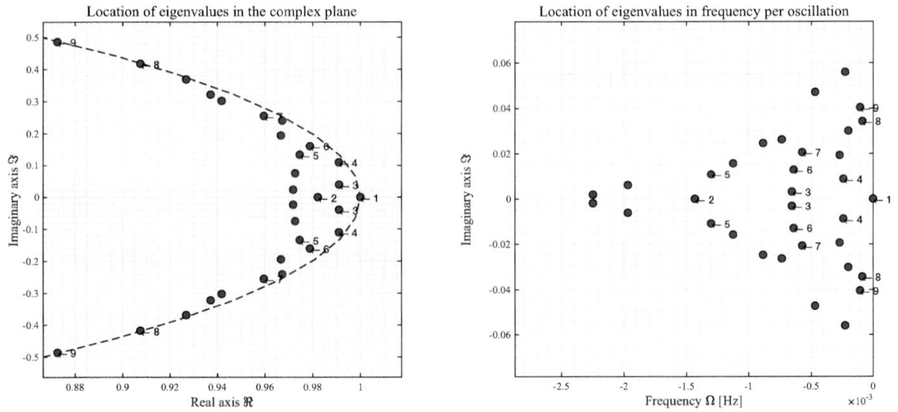

Fig. 6.19 IODMD$_u$ eigenvalues of the ROM for Axial Induction Control. On the left, the discrete time eigenvalues λ on the complex plane are represented relative to the unit circle (dashed line). On the right, the eigenvalues transformed to continuous time are represented. The area where the majority of modes are presented has been zoomed in. Visualization based on the code available in [2]

Each of the modes corresponds to a coherent structure, meaning a spatial pattern of the fluid which evolves in time. The interpretation of these structures follows the same principles as the structures found in the analysis of the ROM for Wake Redirection Control. These coherent structures can be visualized and add value to the interpretation of the dynamics of the wake in collective pitch control, a fundamental step in the further development and usage of state of the art wind farm control strategies, such as dynamic pitch control. Some of the modes have been represented in Figs. 6.20, 6.21, and 6.22, for different frequency domains. It

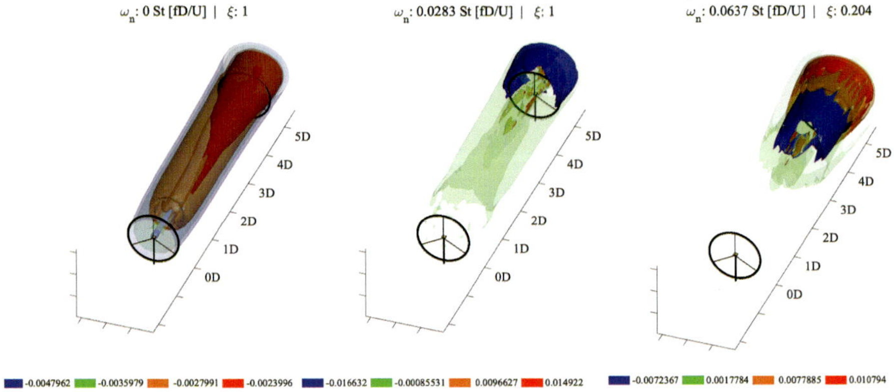

Fig. 6.20 DMD modes #1, #2 and #4. Mode #1 corresponds to a background mode that is not changing (i.e. it has zero eigenvalue). By analysing the isosurfaces, it is seen that this mode represents the deficit velocities always present behind turbine 1. Initial results published in [1]

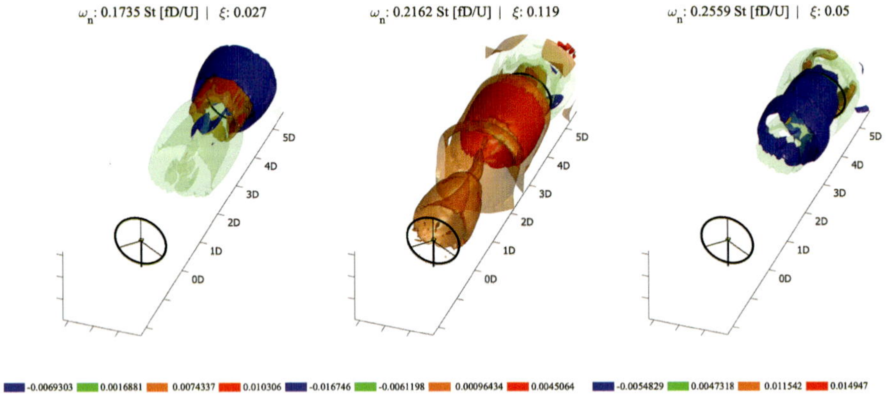

Fig. 6.21 DMD modes #6, #7 and #8, with natural frequencies comprised between 0.17 *St* and 0.26 *St*. The spatial patterns present here appear to also explain some of the dynamics more related with the downstream turbine, as opposite to background and low frequency modes. Initial results published in [1]

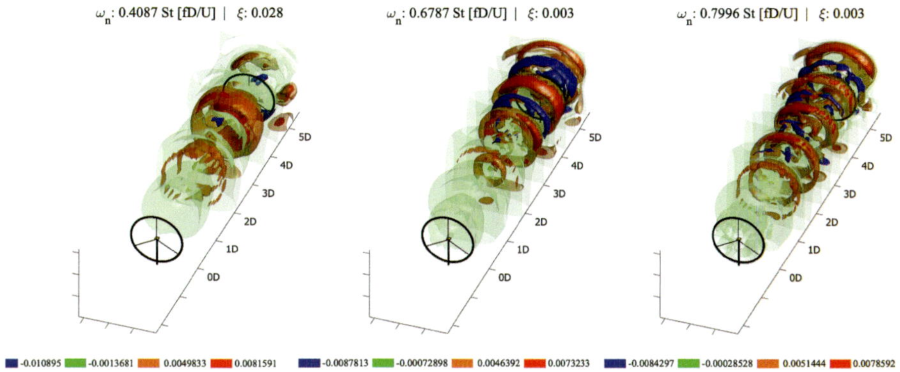

Fig. 6.22 DMD modes #11, #15 and #16, with natural frequencies comprised between 0.41 *St* and 0.80 *St*, showing pulsating structures in-between upstream and downstream turbines. Initial results published in [1]

is important to note that the influence of the control action on the flow field is not represented in these structures. They are merely representations of the high order eigenvectors of state matrix **A**. Isosurface of the different values of the velocity field is sketched so that the complete structure is visualized.

Interpreting dynamic modes can be challenging. Some of these may not necessarily yield direct insight into the evolution of the flow. In addition, as mentioned already, the influence of the control action cannot be evaluated, as these modes are only representations of the state space matrix eigenvectors (projected to the high dimensional space). Furthermore, despite having listed the modes by increasing natural frequency and by the order in which they are identified in the SVD, they may have distinct energetic contributions to the development of the wake, which may still be of interest to investigate.

Despite some of the existing limitations in interpreting the physical significance of all these modes, some interesting patterns can be noticed. For example, in Fig. 6.20, the lower frequency modes are represented. The first one appears to be a background mode, as indicated by the natural frequency of value zero. The colourful structures that can be distinguished all take negative values. This mode most likely contributes to the reconstruction of the always present velocity deficit of the upstream turbine wake. The third mode in Fig. 6.20 depicts structures that are not associated with a deficit in the wake. Colourful structures here represented appear to contribute positively to the reconstruction of the streamwise velocity—apart from the blue structure. This may be related, for example, with increases in the streamwise velocity around the downstream turbine rotor. Comparatively with other modes which have higher frequencies, the damping ratio of this one is high. A possible interpretation is that this particular contribution to the overall flow development is quite stable around this region.

Modes of higher frequency are also of interest to observe, which are represented in Fig. 6.22. These modes are characterized by low damping ratios, indicating increases in the structures seen and associated with rapid growth and decays. Furthermore, pulsating structures, perhaps associated with large vortices in the wake, appear to be encoded in these modes.

6.7 Testing of Non-linear Observables

The streamwise velocity component u was used in the snapshot matrices to compute the ROM. However, there is no guarantee that these represent the best coordinates where the dynamics of interest can evolve. In the modelling of wake redirection due to yaw misalignment, it was seen that using non-linear states can increase the ROM's accuracy. In light of these results, the same is attempted for the modelling of axial induction. Several different states are attempted at first in alternative to using the u velocity component. The results are compiled in Table 6.3.

Table 6.3 Models validation results using IODMD algorithm with different observables. The first column has the data chosen for the states. Second, third and fourth columns have the fit criteria VAF and model size, where VAF of the second turbine generator power is deemed as the most relevant for choosing the most accurate model. The last two columns show the improved performance in terms of VAF (green) and the improvements in terms of the size of model. Reduced models are less computationally expensive for the purposes of model predictive control; hence they are written in green. Initial results published in [1]

Model properties	Model obtained by IODMD_u			Δ IODMD_u	
Observable	$\text{VAF(WT2)}_{\text{max}}$	Model size	$\text{VAF(WT1)}\|_r$	Δ VAF $\text{(WT2)}_{\text{max}}$	Δ size
v	74.78 %	39	97.64 %	+1.19%	+2.63%
w	81.71 %	39	96.99 %	**+8.12%**	**+2.63%**
u'	43.16 %	29	93.99 %	−30.43%	−23.68%
v'	43.09 %	49	96.71 %	−30.50%	+28.95%
w'	55.57 %	49	96.69 %	−18.02%	+28.95%
u'^2	51.83 %	57	95.58 %	−21.76%	+50.00%
w'^2	35.18 %	104	94.78 %	−38.41%	+41.32%
u^2	74.71 %	24	96.76 %	+1.12%	−36.84%
v^2	72.38 %	112	99.33 %	−1.21%	+194.74%
w^2	77.67 %	106	99.29 %	+4.08%	+178.95%
$u \cdot v$	75.56 %	59	99.03 %	+1.97%	+55.26%
$u \cdot w$	76.54 %	35	97.42 %	+2.95%	−7.89%
$v \cdot w$	75 %	120	99.25 %	+1.41%	+63.07%
$u \cdot v \cdot w$	72.83 %	72	99.25 %	−0.76%	+89.47%
$u^2 + w^2$	75.67 %	24	96.79%	+2.08%	−36.84%
$u^2 + v^2 + w^2$	78.65 %	24	96.89%	+5.09%	−36.84%

The attempt to find coordinates which can better reproduce the dynamics of the wind farm system, in light of the Koopman operator theory, appears to be fruitful, showing gains in VAF of 8.13%, when compared to results yielded by IODMD_u. Choosing an observable that captures the complete fluid dynamics, by simultaneously including measurements of all velocity components, such as $u^2 + w^2$ and $u^2 + v^2 + w^2$ also appears to be advantageous, showing increases in VAF of 2.08% and 5.09%, respectively, and decreases in model size, as measured by the total number of states, of − 36.84% for both cases.

6.8 Extended Input Output Dynamic Mode Decomposition with Multiple State Vectors

Leveraging the Koopman operator theory can also be achieved by making use of the EIODMD algorithm, introduced and detailed in the previous Sect. 3.6.2. The EIODMD algorithm consisted in augmenting the state vector with different observables that may be better suited to evolve the existing non-linear dynamics whilst maintaining the baseline pre-existing state. Therefore, returning to the efforts of modelling the wake interaction, the base state u can be augmented with other observables. The snapshot matrices for the EIODMD introduced in Sect. 3.6.2 are now made use of

$$\mathbf{X} = \begin{bmatrix} | & | & & | \\ \boldsymbol{g}(\boldsymbol{x}_1) & \boldsymbol{g}(\boldsymbol{x}_2) & \cdots & \boldsymbol{g}(\boldsymbol{x}_{m-1}) \\ | & | & & | \end{bmatrix} \qquad \mathbf{X}' = \begin{bmatrix} | & | & & | \\ \boldsymbol{g}(\boldsymbol{x}_2) & \boldsymbol{g}(\boldsymbol{x}_3) & \cdots & \boldsymbol{g}(\boldsymbol{x}_m) \\ | & | & & | \end{bmatrix} \tag{6.1}$$

where $\boldsymbol{g} = \left[g_1(x)\ g_2(x)\ \cdots\ g_p(x)\right]^T$. The vector $\boldsymbol{g}$ contains several transformations that can be applied to the data. In theory, all possible functions could be applied to the existing velocity field. Nevertheless, an infinite functional space is not feasible to work with. In order to find a reduced order model, the number of functions applied to the data is finite and given the high dimensionality of the problem (the velocity field is sampled over almost 45 thousand points) it is limited to 3.

The first function $g_1(x) = u$ is chosen to maintain the streamwise velocity as the main state. This is because the ROM will still be used to reconstruct the wake and this has, so far, been the coordinate chosen as the baseline. The remaining functions are then $g_2(v) = v$ and $g_3(x) = w$. The results of augmenting with the different velocity components are summarized in Table 6.4.

The results indicate that augmenting the data matrix with additional information can have benefits. The additional benefits of augmenting u with w information translate in an increase in VAF of 8.15% and a reduction in model size to

Table 6.4 Model validation results using EIODMD algorithm with different observables. The second observables column has the state chosen to augment the initial data matrix with u information. The last two columns represent the average NRMSE of the first observable, using both identification and validation data. Initial results published in [1]

Model properties		Model obtained by EIODMD			Δ IODMD$_u$			
Observable	Second observables	VAF(WT2)$_{max}$	Model size	VAF(WT1)$\|_r$	Δ VAF (WT2)$_{max}$	Δ size	$\overline{\text{NRMSE}}$ (id)	$\overline{\text{NRMSE}}$ (val)
u	v	79.00 %	32	96.70 %	+5.41%	−15.79%	−0.05%	−0.07%
u	w	81.74 %	30	97.27 %	**+8.15%**	**−21.05%**	**−0.12%**	**−0.33%**
u	v, w	79.76 %	28	97.22 %	+6.17%	−2.19%	−0.08%	−0.13%

21.05%. Furthermore, the average NRMSE, which measures that ability of the model to reconstruct the full state information, is reduced to 3.54% and 3.42%, for identification and validation measurements, respectively.

This result should not be of complete surprise. In the last section, the IODMD algorithm was tested with different observables, and it was seen that the coordinate w, as opposed to using u, leads to a model of increased accuracy. Therefore, by creating a state vector with both *u and w*, it is possible to find a ROM with a lower number of state variables that can better capture the dynamics of the downstream turbine power production and still reconstruct the streamwise velocity profile of the wake without decreases in the reconstruction accuracy, when compared to the baseline reconstruction of the IODMD_u model. It is also important to note that the benefit of adding w outweighs the benefit of adding v as a second observable, or even adding both w and v to the initial state vector. This may be explained by the fact that variations in the vertical velocity component may be more suited to describe the captured power dynamics at the downstream turbine.

The $\text{EIODMD}_{u,w}$ ROM predicts the downstream turbine power production with a VAF of 81.74% and reconstructs the wake throughout the simulation with an average NRMSE of 3.42%. The IODMD_u ROM predicts the same signal with a VAF of 73.59% and reconstructs the wake with an average deviation of 3.75%. In addition, the best performing model using the $\text{EIODMD}_{u,w}$ algorithm has 30 states, whereas the chosen IODMD_u one has 38 states. This improvement results from the simple addition of information related to the vertical velocity component w to the snapshot data matrix already containing information regarding the streamwise velocity component u. By performing an addition as opposed to a pure substitution, it is possible to increase the accuracy of both output and state predictions and reconstruct the streamwise velocity deficit, highly important to characterize the wake.

It should be noted that this process can become computationally quite expensive, especially in the case of high order systems. By augmentation, the number of points increases by a factor of 2 from 43.907 to 87.814 data points. Consequently, the SVD is performed on a matrix with a total of almost 80 million entries, as opposed to 40 million.

A comparison of both models' performance is depicted in Fig. 6.23. It is interesting to note how the $\text{EIODMD}_{u,w}$ better captures the spikes in power as a consequence of the blades pitching back to 0 degrees.

Both models can also be compared based on their response in the frequency domain. Another result of interest is the resonance frequency ω_n at approximately 0.2 *St* which is captured and can be observed in the Bode magnitude plot in Fig. 6.24. The system's response at this frequency shows a magnitude of 10.43 dB, higher than any other frequency.

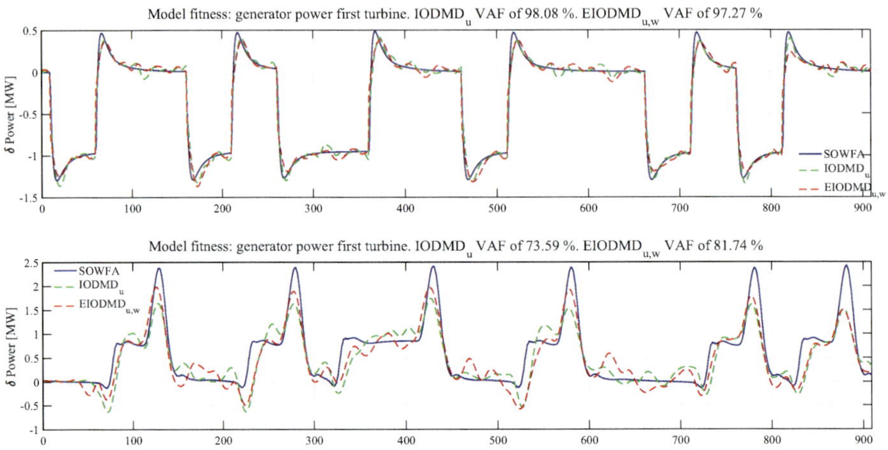

Fig. 6.23 Comparison of IODMD$_u$ and EIODMD$_{u,w}$ model's performance. Initial results published in [1]

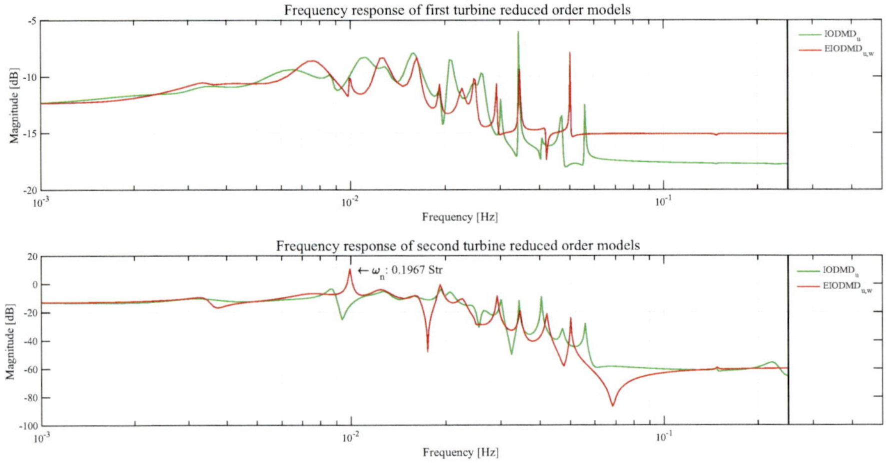

Fig. 6.24 Comparison of IODMD$_u$ and EIODMD$_{u,w}$ model's response in the frequency domain. Initial results published in [1]

6.9 Wake Reconstruction Using Extended Models

In Table 6.4 the improvements achieved with the extended version of the IODMD$_u$ models were summarized. The increase in the VAF of the downstream turbine power production and the decrease in the average NRMSE were emphasized. However, when it comes to the increased accuracy of the wake reconstruction, this is more difficult to grasp. To better understand this, the average NRMSE can be broken down to variations in space and in time. Figures 6.25 and 6.26 depict the evolution of the mean NRMSE for the datasets used for model identification and validation,

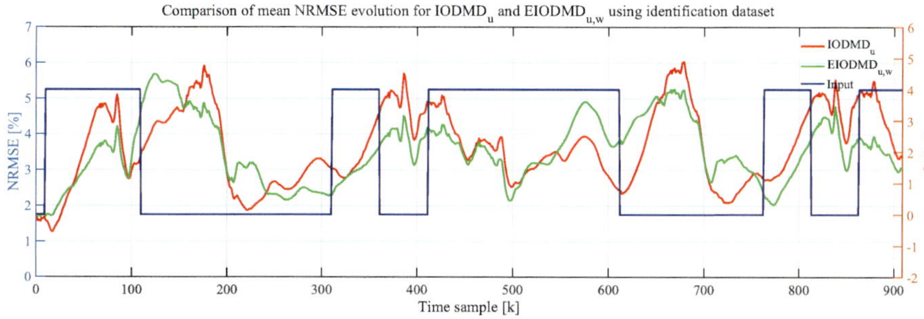

Fig. 6.25 Normalized Root Mean Squared Error (NRMSE) of reconstructed wake using both IODMD_u and $\text{EIODMD}_{u,w}$ Reduced Order Models. Simulation using input data used for identification of models. The mean NRMSE of IODMD_u is shown in red. The mean NRMSE of $\text{EIODMD}_{u,w}$ is shown in green. The input signal, the collective pitch angle of the blades, is shown in dark blue

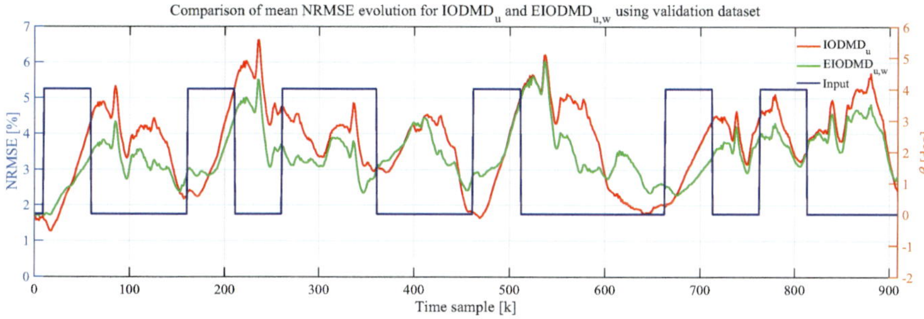

Fig. 6.26 Normalized Root Mean Squared Error of reconstructed wake using both IODMD_u and $\text{EIODMD}_{u,w}$ Reduced Order Models. Simulation using input data used for validation of models. The mean NRMSE of IODMD_u is shown in red. The mean NRMSE of $\text{EIODMD}_{u,w}$ is shown in green. The input signal, the collective pitch angle of the blades, is shown in dark blue

respectively. For the identification dataset, there are moments where the deviations between the reconstructed wake and data from the simulation are higher using the $\text{EIODMD}_{u,w}$ model. However, the case is different for the validation dataset. It can be seen in Fig. 6.26 that the error of the reconstructed wake yielded by $\text{EIODMD}_{u,w}$ is—at the majority of times—smaller than the error of the reconstructed wake yielded by IODMD_u. Although the deviations of the reconstructed wake given by both models follow the same trend, the error of $\text{EIODMD}_{u,w}$ appears to be approximately 1% smaller. This is especially visible at the peak of the deviation, after the blades pitch back to 0 degrees. Time samples in the vicinity of 80 or 250 are examples of this behaviour.

It is not only of interest to evaluate the variation of the deviations of both models with relation to the true flow field not only in time but also in space. Figures 6.27 and 6.28 depict the evolution of the wake for both models, side by side, at hub height. Figures 6.29 and 6.30 compare the wake reconstruction of both models at the downstream rotor plane and further depict the true flow field side by side.

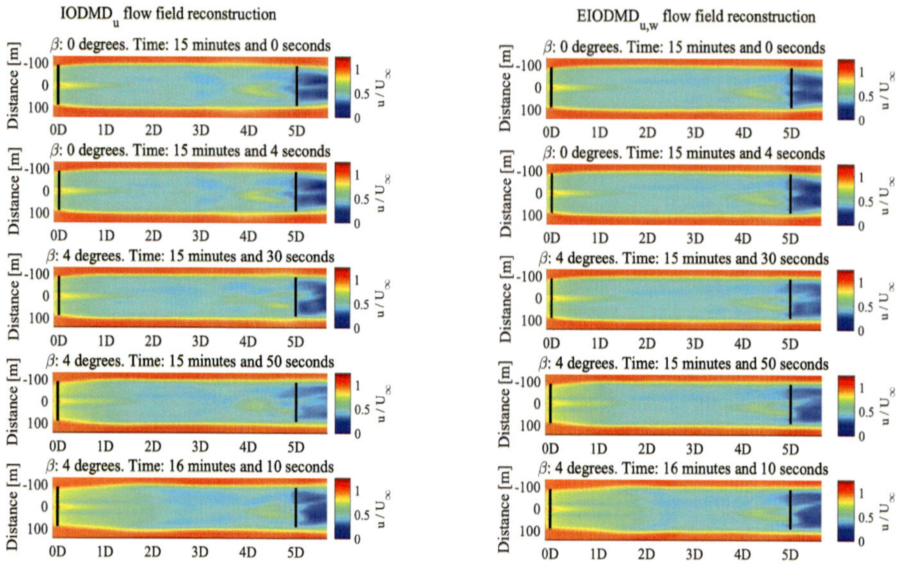

Fig. 6.27 Comparison of flow field reconstruction from low order state trajectory using IODMD$_u$ (left) and EIODMD$_{u,w}$ (right) at hub height. State trajectory obtained using validation data. Key snapshots in time are taken to illustrate the ROM's ability in predicting wake dynamics

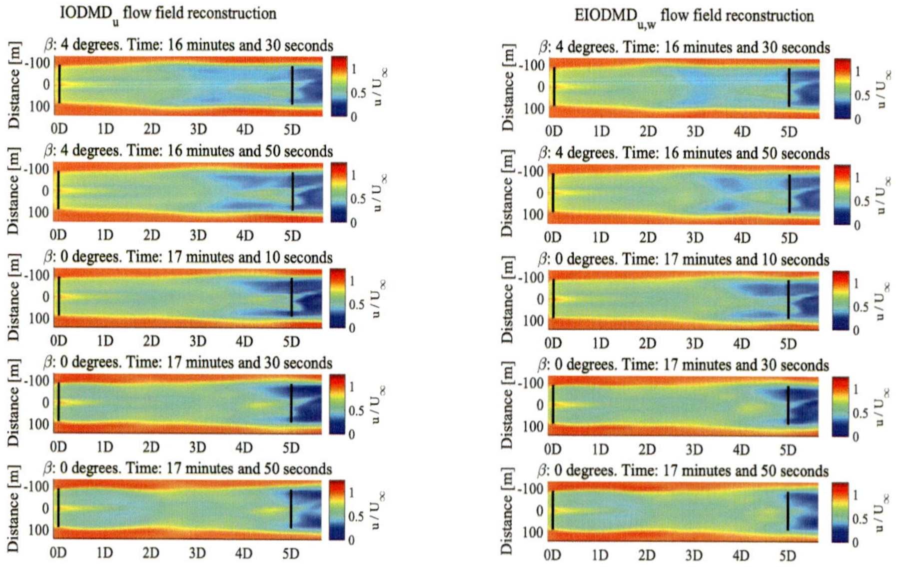

Fig. 6.28 Comparison of flow field reconstruction from low order state trajectory using IODMD$_u$ (left) and EIODMD$_{u,w}$ (right) at hub height. State trajectory obtained using validation data. Key snapshots in time are taken to illustrate the ROM's ability in predicting wake dynamics

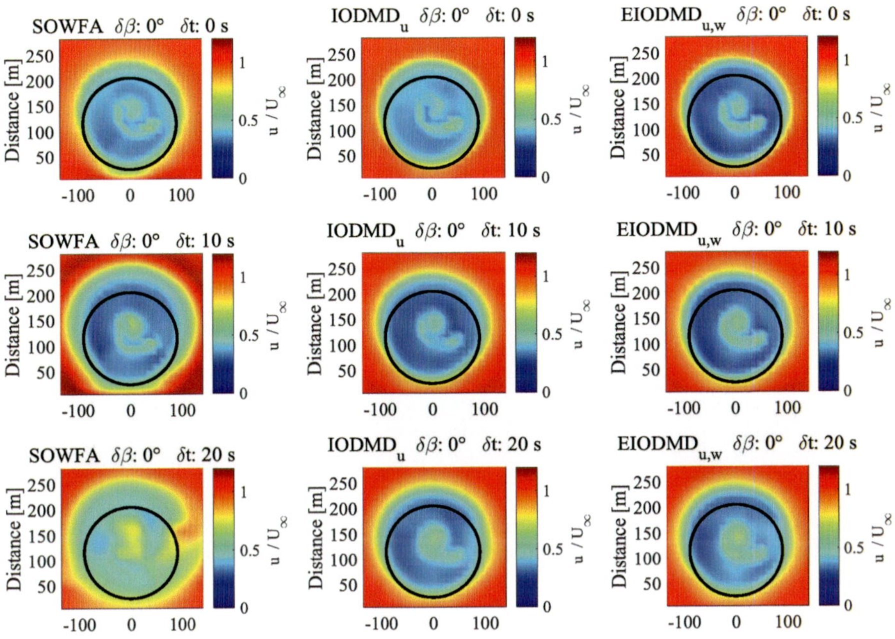

Fig. 6.29 Comparison of SOWFA flow field at downstream turbine rotor plane (left), flow field reconstruction from low order state trajectory using IODMD_u (middle) and $\text{EIODMD}_{u,w}$ (right). State trajectory obtained using validation data

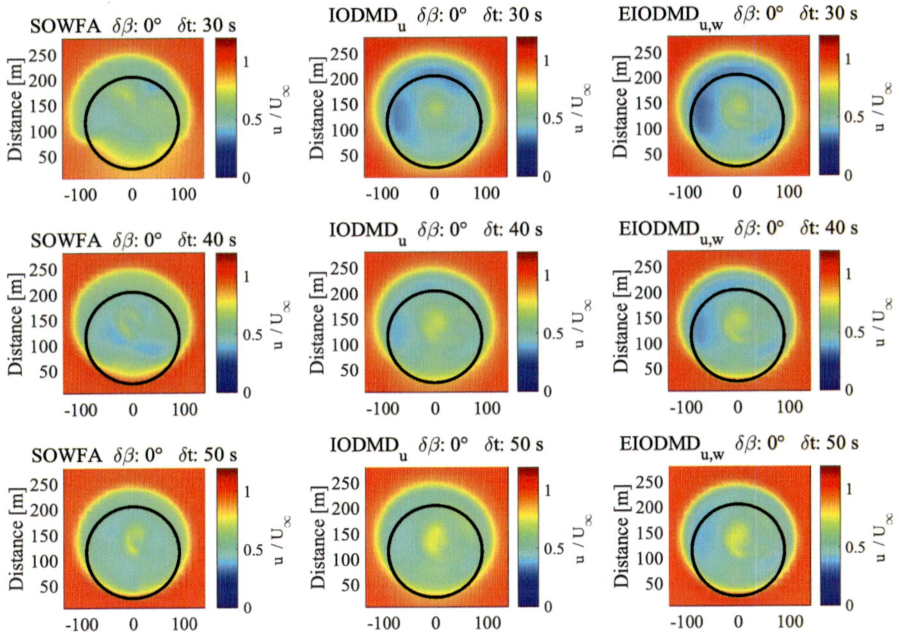

Fig. 6.30 Comparison of SOWFA flow field at downstream turbine rotor plane (left), flow field reconstruction from low order state trajectory using IODMD_u (middle) and $\text{EIODMD}_{u,w}$ (right). State trajectory obtained using validation data

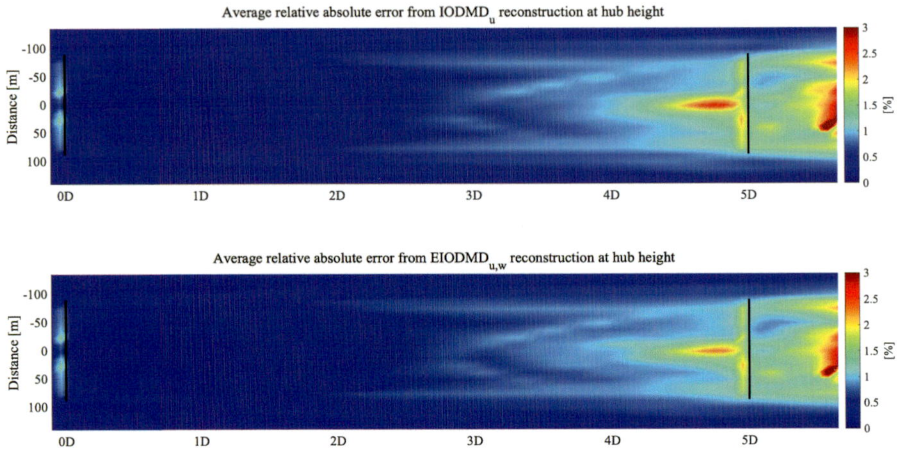

Fig. 6.31 Comparison of the (absolute) time-averaged normalized deviation of the reconstructed wake at hub height. On the upper illustration, the time-averaged deviations of the wake reconstructed by the IODMD_u model with relation to the true flow field are shown. On the lower illustration, the time-averaged deviations of the wake reconstructed by the $\text{EIODMD}_{u,w}$ model with relation to the true flow field are shown

When comparing the reconstructions at hub height, it is interesting to note that in the first snapshots a small depression in the velocity field at 3 rotor diameters appears in the reconstruction by IODMD_u. This is absent in the reconstructed wake using $\text{EIODMD}_{u,w}$. As the blades pitch to 4 degrees, the modified wake is transported downstream, with no difference in particular between both models to be pointed out. Distinct patterns between the reconstructed wakes of both models also appear at the downstream turbine rotor plane. The most prominent feature to be emphasized is the increased velocity region with a curl shape at the centre of the rotor plane. In the last snapshot of Fig. 6.30, for example, it appears this is better reconstructed by the $\text{EIODMD}_{u,w}$ model than the IODMD_u, although there is no guarantee this conclusion is valid for all time instants.

An alternative but complimentary way to gather insight into the ability of the different ROMs in reconstructing the wind turbine wake is to visualize in space the average deviation between the reconstruction and true flow field. In more concrete terms, the average deviation at each point in space is taken for every moment in time, and this is then visualized at hub height and at the downstream rotor plane. The latter is illustrated in Fig. 6.31 and the former in Fig. 6.32. The decreased NRMSE of the reconstructed wake using the $\text{EIODMD}_{u,w}$ model should become evident in these illustrations. In Fig. 6.31, for example, the region near the downstream turbine rotor has a larger deviation when compared to other areas. However, comparatively, this region is lower for the reconstruction using the $\text{EIODMD}_{u,w}$ model. The same should be visible in the downstream turbine rotor plane region, where the deviations between reconstruction and true flow field are smaller in 0.5 to 1% when the $\text{EIODMD}_{u,w}$ model is used.

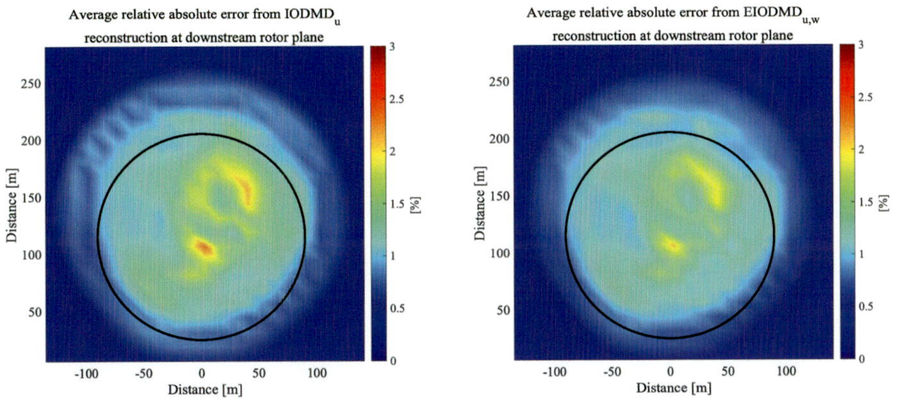

Fig. 6.32 Comparison of the (absolute) time-averaged normalized deviation of the reconstructed wake at the downstream turbine rotor plane. On the left-side illustration, the time-averaged deviations of the wake reconstructed by the IODMD_u model with relation to the true flow field are shown. On the right-side illustration, the time-averaged deviations of the wake reconstructed by the $\text{EIODMD}_{u,w}$ model with relation to the true flow field are shown

It is of interest to compare in more detail the differences between the reconstructed wake yielded by the $\text{EIODMD}_{u,w}$ and the IODMD_u ROM. Using the validation dataset, it is seen that the reconstruction given by the $\text{EIODMD}_{u,w}$ ROM is, at the majority of times, closer to the true flow field. Furthermore, particular patterns in the flow field are seen to be better represented by the later ROM, such as the slight curl shape present at the wake centre at the downstream rotor plane.

6.10 Extended Input Output Dynamic Mode Decomposition with Deterministic Variables

In addition to augmenting the initial data matrices with flow field data, other information related with turbine performance can be experimented with. This follows the logic of the Extended Input-Output Dynamic Mode Decomposition, where the terms $g(x)$ figuring in Eq. 6.1 can also refer to non-flow field information.

Two tests are performed to assess the added value of this variant of the EIODMD algorithm. The first test resembles the approach already employed to calculate the IODMD_u model, where the snapshot matrices are assembled with information of the streamwise velocity component u and further augmented with the so-called deterministic states. The resulting model is referred to as $\text{EIODDMD}_{u,\Omega}$. The rotor speed of the upstream and downstream turbines is chosen as deterministic states.

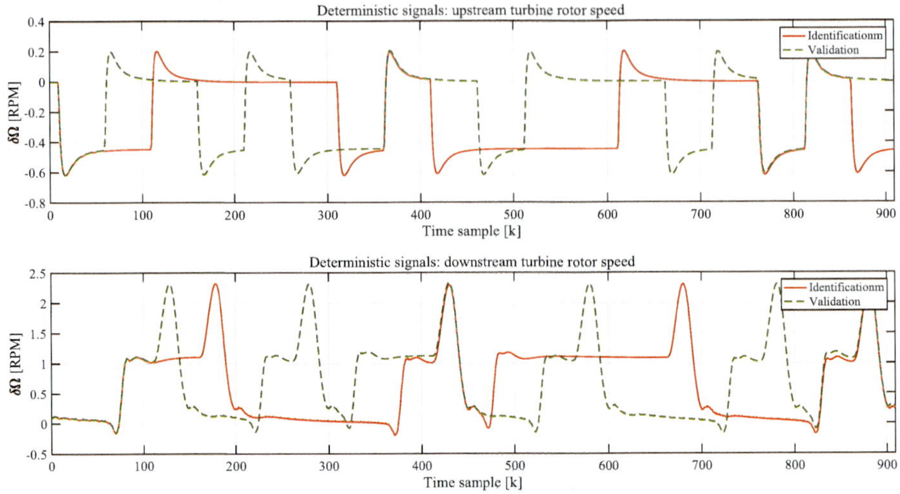

Fig. 6.33 Deterministic signals used for augmenting the state space matrix. On the top plot, the rotor speed of the upstream turbine is shown, for both identification (orange line) and validation (green dashed line). On the lower plot, the rotor speed of the downstream turbine is shown, for both identification (orange line) and validation (green dashed line). Signals are removed from the steady-state average trend

This serves as a mere test, and the reasoning for choosing this also follows the fact that rotor speed measurements are available from commercial turbines or can be estimated resorting to other available signals. This approach should also make possible the reconstruction of the wake, as the ROM could then still be used to estimate the state space trajectory of the streamwise velocity component given a time series of input data. The second test follows the same logic, with the difference that the u information is replaced by w, following the benefits already seen using this approach. The deterministic signals used can be seen in Fig. 6.33. These signals are directly used in the EIODMD algorithm, and the mean value in steady-state operation has already been removed. The mean values removed from the rotor speed signals are 7.20 and 3.67 RPM for the upstream and downstream turbines, respectively.

The results of both these tests are summarized in Table 6.5. The results point to the conclusion that augmenting data matrices with turbine information is a trade-off. For the purposes of prediction and control, it does provide models that are better able to capture the generator power dynamics, compared to the baseline IODMD$_u$ model. This procedure does not incur on an extra computational cost, as only few lines of data are added to the snapshot matrices. Nevertheless, increases in the average NRMSE were registered, to 3.83% and 4.20% for identification and validation, respectively.

Table 6.5 Models validation results using EIODMD algorithm with different base observable and deterministic states. Initial results published in [1]

Model properties		Model obtained by EIODMD			Δ IODMD$_u$			
Observable	Deterministic	VAF(WT2)$_{\max}$	Model size	VAF(WT1)$\|_r$	Δ VAF (WT2)$_{\max}$	Δ size	$\overline{\text{NRMSE}}$ (id)	$\overline{\text{NRMSE}}$ (val)
u	Ω_1, Ω_2	76.71 %	12	98.97 %	+3.12%	−68.42%	+0.17%	+0.45%
w	Ω_1, Ω_2	76.79 %	30	98.93 %	+3.20%	−21.05%	–	–

Another version of the EIODMD algorithm to be tested relies on augmenting the data matrices not with flow field information but with other signals, such as turbine measurements, which have been referred to as deterministic signals. The Singular Value Decomposition is only applied to the velocity field data, making the size of the snapshot data matrices remain practically the same. This approach is therefore less computationally costly. The tests performed show that the VAF, compared to the baseline case yielded by the IODMD_u ROM, increases by an additional 3.12 percentage points. However, the accuracy of the reconstructed wake decreases, with the average NRMSE reducing by 0.45 percentage points.

The performance of the model, measured in terms of the VAF of the downstream turbine power, can be inspected in Fig. 6.34. The VAF is still lower than the $\text{EIODMD}_{u,w}$ ROM, and it can be seen that the jumps surpassing 2 MW in the power production of the downstream turbine are not captured as well. The increase in the average NRMSE is also visible in Figs. 6.35 and 6.36, where the absolute deviation averaged for all instants can be evaluated at hub height and at the downstream rotor plane, respectively. It is shown in Fig. 6.35, for example, that the $\text{EIODMD}_{u,w}$ model is not able to reconstruct the wake between 1 and 4 rotor diameters as precisely as the IODMD_u ROM.

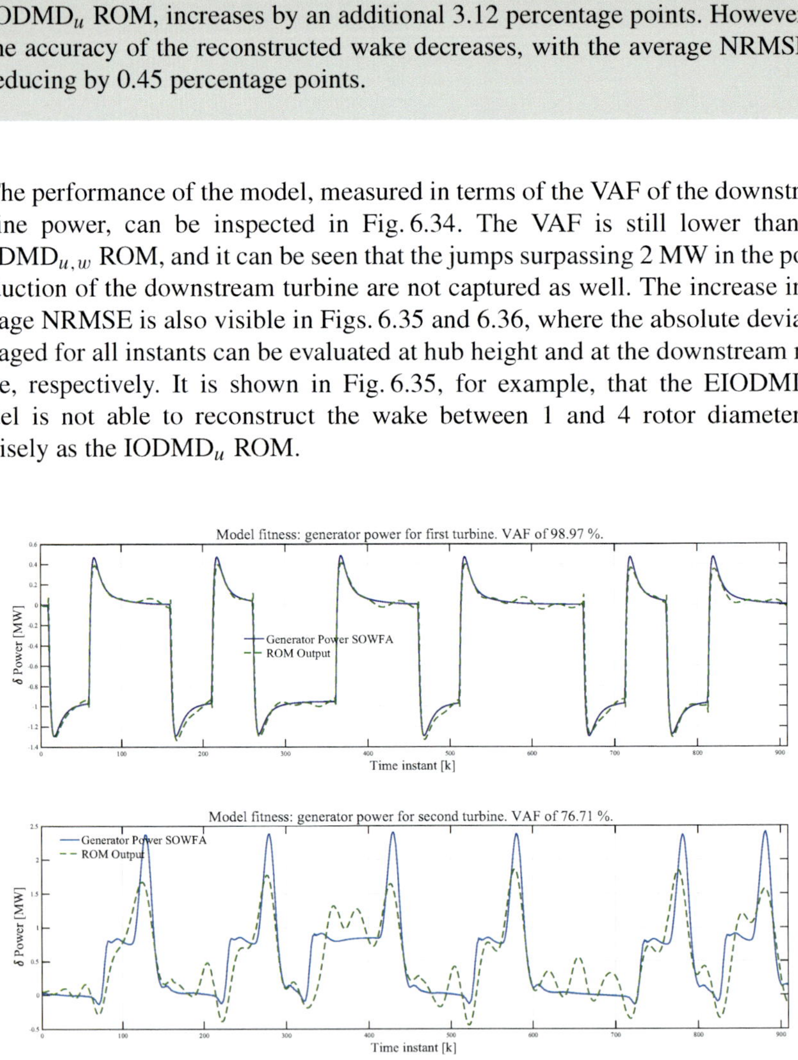

Fig. 6.34 Best performing model validation results for axial induction control] Evaluation of $\text{EIODMD}_{u,\Omega}$ Reduced Order Model performance using validation dataset. Results of linear simulation of ROM with 12 states and comparison with output from SOWFA

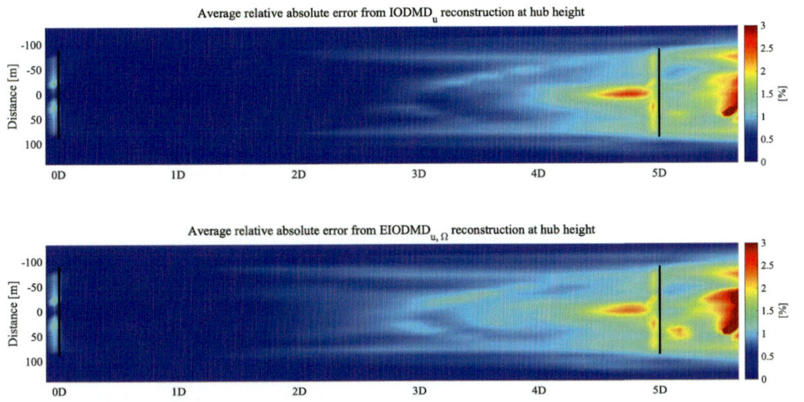

Fig. 6.35 Comparison of the (absolute) time-averaged normalized deviation of the reconstructed wake at hub height. On the upper illustration, the time-averaged deviations of the wake reconstructed by the $IODMD_u$ model with relation to the true flow field are shown. On the lower illustration, the time-averaged deviations of the wake reconstructed by the $EIODMD_{u,\Omega}$ model with relation to the true flow field are shown

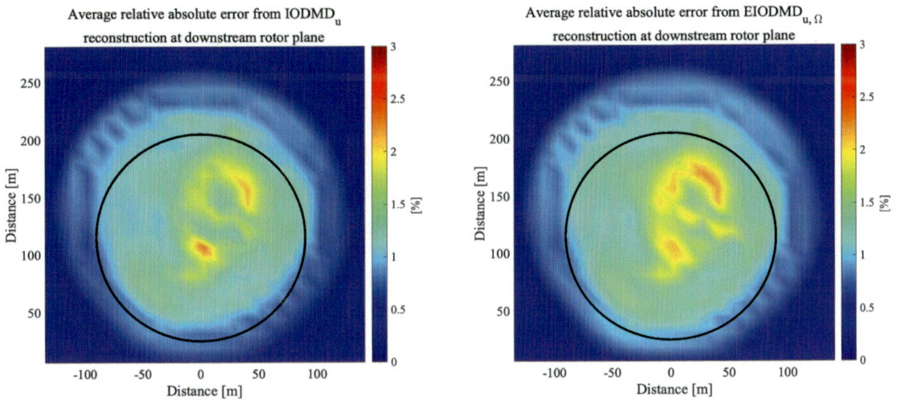

Fig. 6.36 Comparison of the (absolute) time-averaged normalized deviation of the reconstructed wake at the downstream turbine rotor plane. On the left-side illustration, the time-averaged deviations of the wake reconstructed by the $IODMD_u$ model with relation to the true flow field are shown. On the right-side illustration, the time-averaged deviations of the wake reconstructed by the $EIODMD_{u,\Omega}$ model with relation to the true flow field are shown

This chapter exemplified the application of the data-driven techniques introduced in Chap. 3, based on the methodology described in Chap. 4, to the case of Axial Induction Control. The approach taken is the same as in Chap. 5. Based on two datasets gathered from high fidelity simulations, reduced order models of increased sized are calculated and validated. Similarly as in Chap. 5, ideas based on the Koopman operator theory are leveraged in an attempt to better capture the non-linear behaviour in the wake. This proved more useful compared to the case of Wake Redirection Control, as replacing the state information by the vertical velocity component w led to increases in 8% of the Variance Accounted For for the downstream turbine power production. It is also important to note that the dynamics observed in the Axial Induction Control simulations were more challenging to model than the ones in Chap. 5. An additional step taken in this chapter is the testing of the Extended Input Output Dynamic Mode Decomposition approach, where the results gathered showed that augmenting the initial state vector with other variables led to models able to better reproduce the non-linear behaviour. It was showed that augmenting with flow field information, when properly chosen, could both better reproduce the output and reconstruct the wake. However, the task could become computationally expensive as the state matrices double in size. Another option tested was to augment the state matrices with the rotor speed information, referred to as deterministic states, thus decreasing the computational power required when increasing the state matrices. Nevertheless, the benefits were only seen in terms of better capturing output information, not developments in the flow field. Detailed visual comparisons were shown to prove the latter points.

References

1. N. Cassamo, J.-W. van Wingerden, On the potential of reduced order models for wind farm control: a Koopman dynamic mode decomposition approach. Energies **13**(24) (2020). ISSN: 1996-1073
2. J. Nathan Kutz et al., *Dynamic Mode Decomposition: Data-Driven Modelling of Complex Systems* (Siamr, 2016)

Chapter 7
Advanced Topics

In the previous chapters, the system identification loop was followed for two different flow control strategies. Using two datasets for each strategy, one for identification and another for validation, Reduced Order Models (ROMs) were computed and validated. This process used several tools, such as inbuilt MATLAB functions, which have their own underlying assumptions. The first aim of this chapter is to discuss such underlying assumptions, more specifically the initial conditions used in the linear simulations and the methods used to estimate the latter. These will be explored, along with the implications of using different methods. Furthermore, the models were used mainly for the first two applications of data-driven models discussed in Chap. 3: to evaluate underlying dominant spatio-temporal patterns (diagnostics) and to use these to build models capable of predicting the system future behaviour (future state prediction). However, for more ambitious applications, such as control, other requirements may take place. For example, the ROM must be stable. For other applications, such as Model Predictive Control (MPC), a certain degree of causality may make it simpler to formulate the MPC problem mathematically.[1] The methods discussed in this book do not ensure that the final ROM is stable, for example. There are, nevertheless, techniques capable of circumventing this challenge (up to a certain degree). This chapter also discusses these techniques, which were already introduced in Sect. 3.7, and uses examples to discuss its implications. Because these topics are related to the nuances of modelling and simulation, these are referred to as *advanced topics*.

[1] A model is said to be causal when its output depends on future states based on the current states and inputs and therefore does not depend on future ones.

N. Cassamo, J.-W. van Wingerden, *Data-driven Modelling of Wind Farm Flow Control Strategies*, https://doi.org/10.1007/978-3-031-84116-3_7

7.1 Estimation of Initial Conditions

In the previous sections, results concerning the performance of different ROMs were presented, explained and discussed. Two metrics were used to evaluate the performance of the ROMs. The first was the Variance Accounted For (VAF) and the second the Normalized Root Mean Squared Error (NRMSE). The first was used to evaluate the ROM capability to predict the desired outputs, in this case, to predict the power generated by the upstream and downstream turbines. The NRMSE, on the other hand, was used to evaluate the models' capabilities to reconstruct the upstream turbine wake. Both were obtained directly from a linear simulation of the state space model, with the wake being reconstructed by projecting the state space trajectory from the low dimensional space to the high dimensional space. This transformation of coordinates was possible by using the left singular vectors contained in the matrix U obtained from the Singular Value Decomposition (SVD) of the snapshot matrix.

The focus of this subsection is to discuss the details surrounding the linear simulations performed, especially concerning the estimation of initial conditions. Initial conditions refer to the starting values of the system's state at the beginning of a simulation. These will be determining in how the system will respond over time. In many engineering applications, for example, the behaviour of many systems is defined by differential equations. The solution of these equations specifies the system behaviour and response. The initial conditions, therefore, may be required to fully define the response of a system. Consider the classical example of a mass-spring system, where the second-order linear differential equation in 7.1 describes the motion of the mass given a mass m attached to a spring with constant k:

$$m\frac{d^2x}{dt^2} + kx = 0 \tag{7.1}$$

The general solution of the system is written in Eq. 7.3, considering that Eq. 7.1 can be rewritten to the format in 7.2, where $\omega_0 = \sqrt{\frac{k}{m}}$ is the natural angular frequency of the system.

$$\frac{d^2x}{dt^2} + \omega_0^2 x = 0 \tag{7.2}$$

$$x(t) = A\cos(\omega_0 t) + B\sin(\omega_0 t) \tag{7.3}$$

In Eq. 7.3, A and B are constants determined by the initial conditions of the problem. By taking the former equation and its derivative, so as to obtain the expression for the velocity $v(t)$, and evaluating both equations at $t = 0$, A and B are respectively, x_0 and v_0/ω_0, and Eq. 7.3 can be rewritten as

$$x(t) = x_0\cos(\omega_0 t) + \frac{v_0}{\omega_0}\sin(\omega_0 t) \tag{7.4}$$

It therefore becomes evident by the final expression dictating the mass-spring position behaviour over time in 7.4 the relevance of initial conditions. The initial position and initial velocity, as could very well be expected, determine the evolution of the system position by contributing to the sine and cosine components of the sinusoidal motion.

Juxtaposing to the case of the ROM and its state space trajectory, the evolution of the wake through time is not analytically known resorting to the underlying differential equations. As we have seen, the data-driven approach is used to estimate a ROM which embeds the complex system dynamics. Nevertheless, the full behaviour of the wake through time also requires information of initial conditions, which are used when simulating the state space system. How these initial conditions are estimated and the implications of using well-known methods are discussed in the renaming part of this subsection.

As briefly touched upon in the previous chapter, the $lsim$ function simulates the time response of a certain dynamic system given a certain input signal and initial conditions. The dynamical system in this work is the ROM and the input signals the control signal used, depending on the wind farm control strategy being studied. The initial conditions, if omitted, reduce to zero. For the cases explored in previous sections, the initial conditions are known to be not zero, and these correspond to the state vector in the low-dimensional space at the begging of the simulation.

There are two approaches that could be employed to estimate the initial conditions. The first consists in projecting the complete flow field information at the initial time instant to the low order dimensional space by making use of the left singular vectors contained in U. This was the approach taken for the results previously shown. Examples will be provided in the remaining paragraphs to exemplify this process. The second method consists in estimating the initial conditions of the system by maximizing the fit between the outputs obtained from the validation dataset and the model's response. This can be performed by using an inbuilt MATLAB function such as `findstates`. The initial conditions of the state space ROM are therefore estimated directly in the low order dimensional space. However, there is no guarantee that the estimated initial conditions following the second approach lead to any representative flow field. It is important to note that initial conditions should have some physical relevance; otherwise predictions given by the model may not be realistic. Going back to the example of the mass-spring system, if initial conditions do not reflect a realistic and physically meaningful initial position and velocity, the prediction given by the solution to the differential equation in 7.4 is of no use. This section explores in more detail this dichotomy. It is seen that despite the VAF of the ROM not being sensitive to the initial conditions, the reconstruction of the wake is.

To illustrate the duality briefly mentioned, an example is provided in Listing 7.1. The example starts by importing the results of the IODMD$_u$ model and selects the model with 38 states. This is the best performing model used for Axial Induction Control. The example proceeds by estimating the models' initial conditions using the two methodologies introduced: $\texttt{xo}_{\texttt{est}}$ corresponds to the initial conditions estimated by projecting the flow field in the first time instant from the validation dataset to the low dimensional space by making use of U. The initial state vector $\texttt{xo}_{\texttt{opt}}$, on the other hand, is estimated by maximizing the fit between output signals from the validation dataset and the ROM output using the input signal from the validation dataset. The function `dinit` is part of the LTI Systems Identification Toolbox from the Delft Center of Systems and Control and performs the same as MATLAB's inbuilt function `findstates`. Two linear simulations are then performed, each one using the different initial conditions. The ROM output signals are extracted along with the state space trajectories from each simulation. The VAF of both outputs is assessed and the different wake evolutions through time reconstructed from the state space trajectories. The NRMSE for each one is then assessed.

Listing 7.1 Linear simulation of ROM considering different sets of initial conditions

```
rom_u=load('../IODMD_pitch_u/DMDresults_IODMD_u/RESULTS.mat')
model=rom_u
si=38;

%xo being estimated with projection
xo_est=model.U(1:size(model.states,1),1:si)'*model.statesvalid(:,1);

%xo being estimated by maximizing fit
xo_opt=dinit(model.sys_red{si}.A,model.sys_red{si}.B,model.sys_red{si}.C,model.
     sys_red{si}.D,[model.Inputs_val]',[model.Outputs_val]');

%linear simulations with initial different conditions
[ysim_est, t, xout_est]=lsim(model.sys_red{si}, [model.Inputs_val]',[],xo_est);
[ysim_opt, t, xout_opt]=lsim(model.sys_red{si}, [model.Inputs_val]',[],xo_opt);

%VAF for each of the outputs in the simulation
FITje_est=vaf([model.Outputs_val]',ysim_est);
FITje_opt=vaf([model.Outputs_val]',ysim_opt);

recon_est=model.U(1:n,1:si)*xout_est';
recon_opt=model.U(1:n,1:si)*xout_opt';

xo_est_ho=model.U(1:n,1:si)*xo_est;
xo_opt_ho=model.U(1:n,1:si)*xo_opt;
```

What is seen in the results is that the initial conditions estimated by both methods do not lead to the same flow field representation and that this will affect the wake evolution reconstruction. $\texttt{xo}_{\texttt{est}}$ represents indeed the initial flow field. This is however expected as it was computed directly by projection of the initial flow field from the validation data. $\texttt{xo}_{\texttt{opt}}$, on the other hand, when written in the high dimensional coordinate system, shows little resemblance to the initial wake conditions from the validation data. This can be seen by inspecting the wake reconstruction at the initial time instant by making use of both methods, which is represented at hub height in Fig. 7.1, and at the downstream rotor plane in Fig. 7.2. It becomes therefore clear that the estimation of the initial conditions through methods commonly used in the systems identification literature might not be suited for grey box models where the initial conditions and state space trajectory are of relevance to the final user.

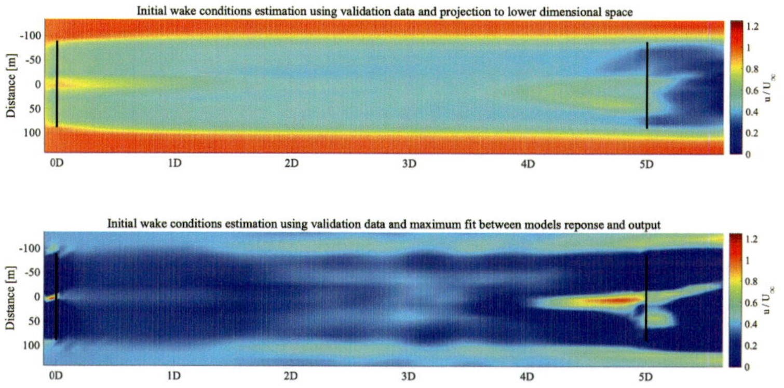

Fig. 7.1 Representation of Reduced Order Model initial conditions state vector in the high dimensional coordinate system at hub height. On the top illustration, the wake reconstruction at hub height using the initial conditions state vector $\mathtt{xo}_{\mathtt{est}}$ estimated directly from the flow field information. On the lower illustration, the initial wake reconstructed from the low order state vector $\mathtt{xo}_{\mathtt{opt}}$ estimated via fit maximization between input and output data using ROM state space matrices. ROM with 38 states used

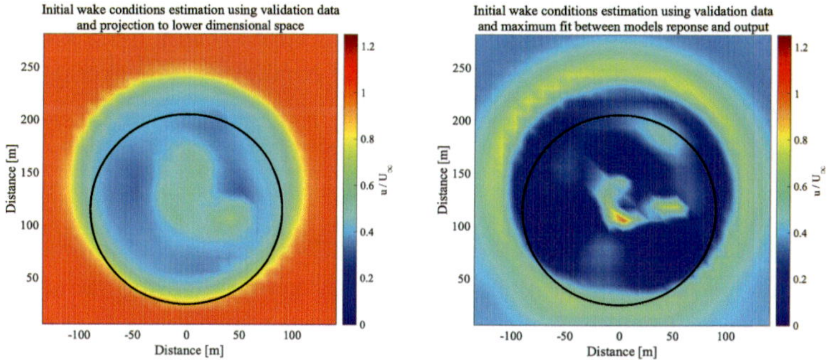

Fig. 7.2 Representation of Reduced Order Model initial conditions state vector in the high dimensional coordinate system at the downstream rotor plane. On the left illustration, the wake reconstruction at the downstream rotor plane using the initial conditions state vector $\mathtt{xo}_{\mathtt{est}}$ estimated directly from the flow field information. On the right, initial wake reconstructed from the low order state vector $\mathtt{xo}_{\mathtt{opt}}$ estimated via fit maximization between input and output data using ROM state space matrices. ROM with 38 states used

Two methods can be employed to estimate the initial conditions to be used in the linear simulations for model validation. The first projects the flow field in the initial time instant to the ROM state space, and the second maximizes the fit between the desired output and validation data. It is seen that the initial conditions estimated by both methods may differ substantially. Furthermore, it is seen that a proper estimation of the initial conditions is fundamental so

(continued)

that the state space trajectory that follows accurately represents the changes taking place in the wake.

In addition, it is curious to note that the same pattern in the wake can be seen despite the unphysical reconstruction. In other words, it appears that particular patterns of the streamwise velocity near the downstream turbine rotor—visible at the centre of Fig. 7.1 from 4D to 5D—or in the outer region of the rotor plane in Fig. 7.2 are still present. These patterns are also observable in the real flow field, however assuming other values. This could be partly explained by the fact that there may exist multiple representations of the flow field in the low dimensional state space vector that are still able to reproduce the generator power dynamics.

It is important to emphasize the duality of the ROM performance when simulated with different initial conditions: The VAF of the downstream generator power is practically the same in both linear simulations, and however the error of the reconstructed wake is not. In Fig. 7.3 it can be seen that the NRMSE of the reconstructed wake when the simulation is performed with $\texttt{xo}_{\texttt{opt}}$ as initial conditions is much higher throughout the complete simulation when compared with the NRMSE of the reconstructed wake when $\texttt{xo}_{\texttt{est}}$ is used as initial conditions. The average NRMSE for the first simulation is of 55.74% whereas for the second is 3.75%. By contrast, the VAF of the second output signal, the downstream generator power, is 73.59% for the simulation using $\texttt{xo}_{\texttt{est}}$ and for the one using $\texttt{xo}_{\texttt{opt}}$ as initial conditions 76.91%. This can be observed in greater detail in Fig. 7.4.

The ROM used for the above simulations has 38 states. In case another ROM is used with a different number of states, other trends are of interest to see and discuss. To exemplify the different trends present, the same example as before is performed; however at this time a ROM with 90 states is used.

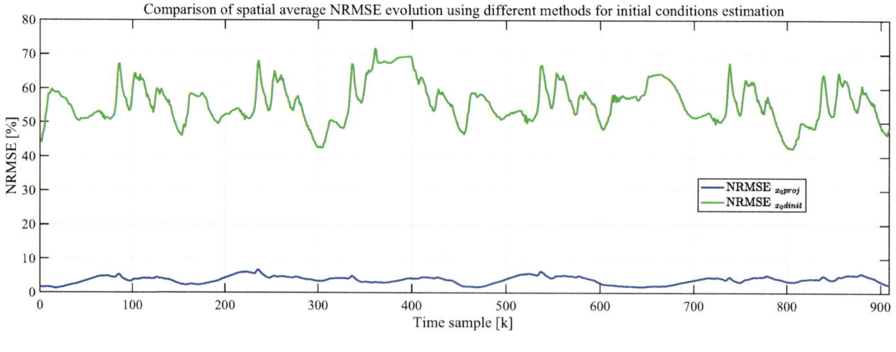

Fig. 7.3 Normalized Root Mean Squared Error of wake evolution reconstructed from the state space trajectory. Represented as blue and green lines are the NRMSEs of the wake reconstructions using the state space trajectory retrieved from the two linear simulations using $\texttt{xo}_{\texttt{est}}$ and $\texttt{xo}_{\texttt{opt}}$, respectively. ROM with 38 states used

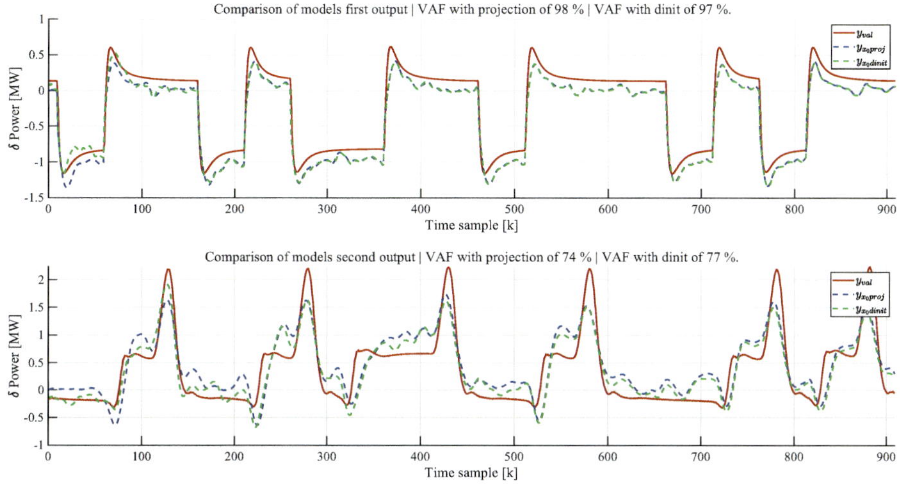

Fig. 7.4 Reduced Order Model outputs for linear simulations using different initial conditions. Upper plot corresponds to first turbine generator power and lower plot to downstream turbine generator power. Output validation data represented with an orange line. Linear simulation output using xo_{est} represented with blue dashed line and using xo_{opt} represented with green dashed line. ROM with 38 states used

Table 7.1 VAF and NRMSE using initial conditions xo_{est} and xo_{opt} with 39 states ROM

	VAF [%]	NRMSE [%]
xo_{est}	73.59	3.75
xo_{opt}	76.91	55.74

Table 7.2 VAF and NRMSE using initial conditions xo_{est} and xo_{opt} with 90 states ROM

	VAF [%]	NRMSE [%]
xo_{est}	55.40	4.25
xo_{opt}	73.75	29.42

In this second example, the VAF of the second output—power produced at the downstream turbine—as predicted by the ROM is of 55.40% using xo_{est} and 73.75% using xo_{opt}, lower than the ROM with 38 states. These metrics for both models and initial conditions are summarized in Tables 7.1 and 7.2.

The average NRMSE of the reconstructed wake is of 4.25% when xo_{est} is used for the linear simulation, whereas it is of 29.42% for the case where xo_{opt} is used. It is interesting to see that when xo_{opt} is used the VAF of the downstream turbine generator power for the ROM with 90 states decreases from 76.91% to 73.75% when compared with the ROM with the 38 states, whereas the average NRMSE decreases from 55.74% to 29.42%. When this is evaluated in more detail, it is seen that despite the estimated initial conditions deviating more from the true flow field—the ROM with 90 states estimates an initial condition xo_{opt} which deviates[2] from the wake

[2] Metric computed by taking an average of the relative deviation with relation to the true flow field over all grid points where the wake is reconstructed.

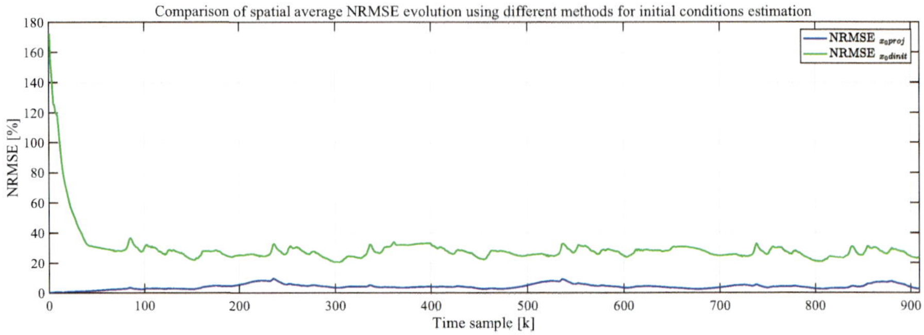

Fig. 7.5 Normalized Root Mean Squared Error of wake evolution reconstructed from the state space trajectory. Represented as blue and green lines are the NRMSEs of the wake reconstructions using the state space trajectory retrieved from the two linear simulations using $\mathtt{xo}_{\mathtt{est}}$ and $\mathtt{xo}_{\mathtt{opt}}$, respectively. Predictions obtained using the 90 states ROM

by 172.61%, whereas the ROM with 38 states deviates by 44.91%—the state space trajectory converges to a flow field which deviates from the validation data by much less.

The latter convergence can be seen in Fig. 7.5. Comparing the wake reconstruction error predicted by the 90 state ROM with the evolution of the reconstructed wake NRMSE given by the ROM with 38 states in Fig. 7.3, it can be seen that 100 seconds after the simulation has started, the NRMSE converges and is bounded between 20% and 30%.

It is also of interest to inspect in more detail and see how the wake is reconstructed and evolves over time for the simulation where $\mathtt{xo}_{\mathtt{opt}}$ is used. This is illustrated in Fig. 7.6. The modification of the wakes' characteristics as the axial induction factor is modified due to the collective motion of the blades pitching is still represented in the state space trajectory. The decreased wake velocity deficit propagating downstream is predicted and well represented. However, it is as the reconstructed wake still differs from the true flow field by a certain offset.

7.2 Regularization

Using the IODMD algorithm and its variants, there is no guarantee that the state space ROM will be stable. However, techniques from systems identification can be used as a post-processing step in order to ensure stability of the model. The technique used to achieve this was briefly explained in Sect. 3.7. This subsection dives deeper and illustrates it with an actual example.

Before diving into the details, let us first clarify the concepts of regularization and stability, briefly mentioned in previous chapters. Regularization refers to techniques used to improve the final model. When using data to identify a dynamical system,

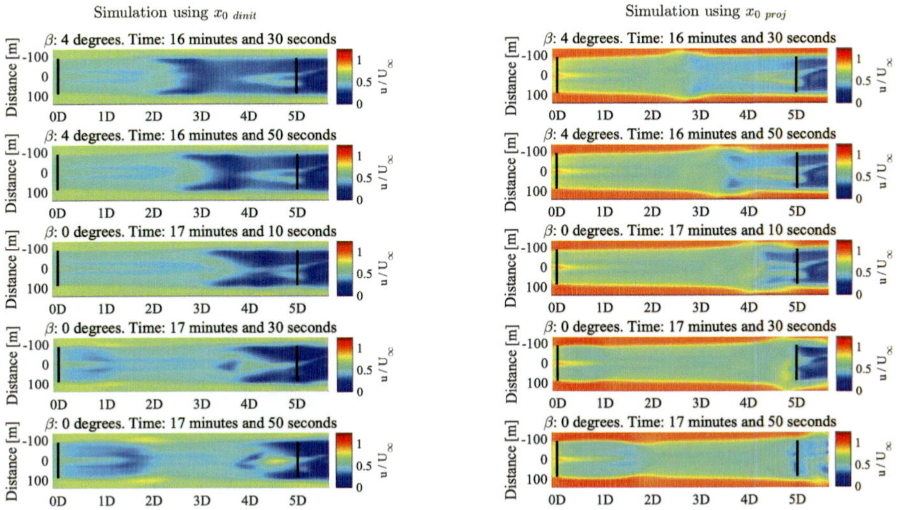

Fig. 7.6 Comparison of wake reconstruction throughout the simulation given by the ROM state space trajectory simulated using the IODMD$_u$ model with 90 states. On the left, wake reconstructed from simulation using `xo`$_{\texttt{est}}$ and on the right using `xo`$_{\texttt{opt}}$. Predictions obtained using the 90 states ROM

the user may penalize certain model parameters based on knowledge of the data to improve the model. Consider, for example, a linear regression model. This kind of statistical model assumes a linear relation between observations of a certain dependent variable y_i to the so-called regressors (or explanatory variables) $x_{i,j}$ through a set of parameters or coefficients β_j and an error term ϵ_i which captures the effect in the output y_i not fully encapsulated by the regressors. This relation can be written as described in the equation below:

$$y_i = \beta_0 + \beta_1 x_{i,1} + \beta_2 x_{i,2} + \ldots + \beta_p x_{i,p} + \epsilon_i \tag{7.5}$$

Ordinary Least Squares (OLS) is a method used to estimate the set of parameters β_p by minimizing the cost function J penalizing the difference between the output y_i and the predicted value $\hat{y}_i$ as depicted in Eq. 7.6. The equation can also be more elegantly written in the format in 7.7.

$$J(\beta_0, \beta_1, \ldots, \beta_p) = \sum_{i=1}^{n} (y_i - (\beta_0 + \beta_1 x_{i,1} + \beta_2 x_{i,2} + \ldots + \beta_p x_{i,p}))^2 \tag{7.6}$$

$$J = \sum_{i=1}^{n} (y_i - \sum_{j=1}^{p} x_{i,j} \beta_j)^2 \tag{7.7}$$

Typical regularization methods consist in adding a penalty term to the cost function in 7.7. Well-known methods include, for example, the Lasso regression ($l1$ regularization) and the Ridge regression ($l2$ regularization). The Lasso regression adds a penalty term consisting of the sum of the absolute value of the coefficients, controlled by the λ parameter, as depicted in 7.8. The Ridge regression adds a penalty term consisting of the squared sum of the coefficients, as equated in 7.9.

$$J = \sum_{i=1}^{n}(y_i - \sum_{j=1}^{p} x_{i,j}\beta_j)^2 + \lambda \sum_{j=1}^{p} |\beta_j| \tag{7.8}$$

$$J = \sum_{i=1}^{n}(y_i - \sum_{j=1}^{p} x_{i,j}\beta_j)^2 + \lambda \sum_{j=1}^{p} \beta_j^2 \tag{7.9}$$

By penalizing the sum of the coefficients absolute values, Lasso regression promotes the reduction of some of the coefficients to 0, thus reducing the complexity of the model by removing parameters that are less important in predicting the output. Ridge regression promotes the reduction of all coefficients by an amount proportional to their values.

In Sect. 3.7 a mathematical derivation to post-process ROMs based on a regularization which ensures model stability is derived, which will be exemplified in this chapter. The example in the previous paragraphs relied on a linear regression model. An OLS method was used to derive the coefficients of the model. This can also be applied to the ROM state space matrices being computed in our examples. In [1], a new and specific type of regularization term is added to the least squares formulation in 7.10, as explained in Eq. 3.56. The regularization term added is the trace[3] of the product of the system matrix $\hat{\mathbf{A}}$, a positive-semidefinite matrix $\mathbf{W}$ and the transpose of the system matrix $\hat{\mathbf{A}}^T$, i.e. $Tr(\hat{\mathbf{A}}\mathbf{W}\hat{\mathbf{A}}^T)$.

$$\min_{\hat{\mathbf{A}},\hat{\mathbf{B}},\hat{\mathbf{C}},\hat{\mathbf{D}}} \left\| \begin{bmatrix} \hat{\mathbf{X}}_{i+1} \\ \mathbf{Y}_{i|i} \end{bmatrix} - \begin{bmatrix} \hat{\mathbf{A}} & \hat{\mathbf{B}} \\ \hat{\mathbf{C}} & \hat{\mathbf{D}} \end{bmatrix} \begin{bmatrix} \hat{\mathbf{X}}_i \\ \mathbf{U}_{i|i} \end{bmatrix} \right\|_F^2 \tag{7.10}$$

This regularization term added to the cost function imposes stability of the least squares solution. The authors in [1] choose $\mathbf{W}$ as the identity matrix. This approach yielded better results, and the authors also motivate this choice from regularization theory through the entropy principle, where all directions are equally penalized.

Stability is desired in any model. However, identifying a model from a finite number of data points which may contain high noise levels, the least squares solution may yield an unstable system. This is possible despite the true underlying system being stable. A discrete time system is unstable if one of the poles lies outside the unit circle (or, in continuous time, the real part of one of the poles is positive). If that

[3] The trace of a matrix is the sum of its diagonal elements.

is the case, for a given bounded input, the output of the model will grow infinitely and not be bounded. For the sole purpose of using DMD related algorithms, there is nothing inherently erroneous in having an unstable model, where most likely one of the modes can be interpreted as instability in the flow or a certain coherent structure that will grow as time passes and contribute to the overall evolution of the flow by a growth in energy. However, for control applications, it may be desired that there are no instabilities in the model. For example, if a Model Predictive Controller is designed based on the ROM, as in the work in [2], the output of the model over a certain prediction horizon should not be unbounded.

The remaining part of the chapter focuses on applying the regularization technique to stabilize the ROM as a post-processing step. The first step to take consists in evaluating which of the IODMD$_u$ ROMs are stable and which are not. In this example, we evaluate the stability of all ROMs by making use of the *isstable* command. This is a MATLAB inbuilt command which checks if a system is stable or not and yields the Boolean corresponding value. The result is illustrated in Fig. 7.7. By visual inspection, it appears that not all ROMs are stable.

However, from Fig. 7.7 it is not possible to see just how far from the unit circle the poles will lie. The model with 16 singular values is unstable, and this is further investigated in this example. In Fig. 7.8 the poles of this ROM can be seen. On the left side of the picture it can be seen the poles are located very close to the unit circle, indicating very slow dynamics. On the right side a zoomed in perspective is provided. Two black arrows point towards the pair of conjugate poles that lie outside the circle and cause the model to be unstable.

The regularization technique introduced in Sect. 3.7 is used to ensure model stability. The entries of the positive definite matrix **W** are set to 100, and the positive

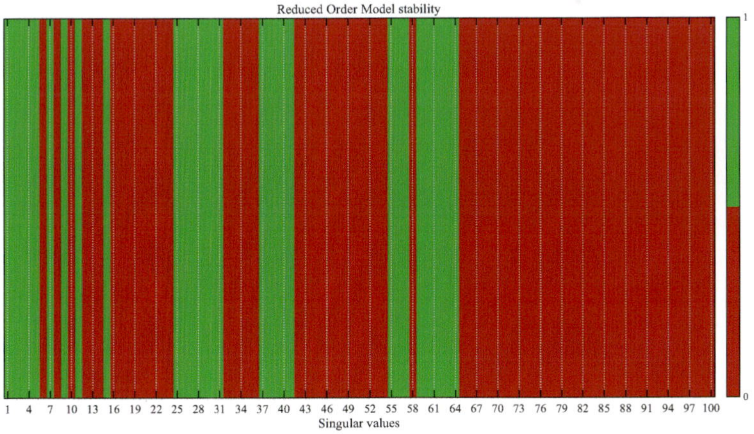

Fig. 7.7 Stability of the different Reduced Order Models based on the output of inbuilt MATLAB function *isstable*. Stable models are represented with a green colour, referring to the output *True*. Unstable models are represented with a red colour, referring to the output *False*

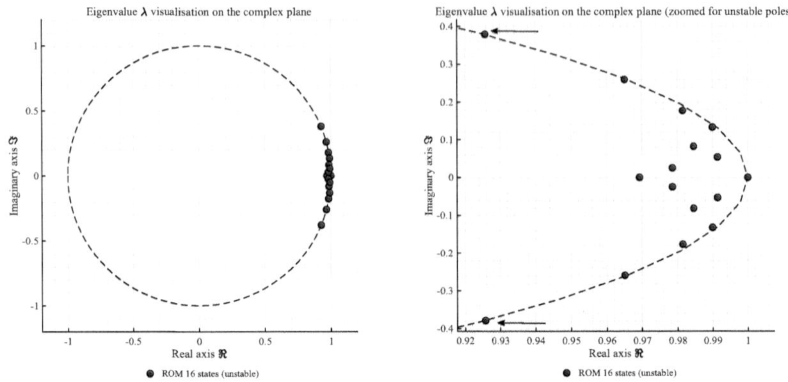

Fig. 7.8 Location of ROM (with 16 singular values) poles. On the left side, the relative location within the complete unit circle can be observed. On the right side, a zoomed in version of the same locations is illustrated, where the black arrow represents the poles that lead to an unstable response of this ROM

real scalar c is varied. Different values are tested, namely 10, 20, and 40. This can be seen in detail in the code snippet in Listing 7.2.

Listing 7.2 Example of ROM post processing regularization step

```
[model_regul_10]=regul(model_chosen, 16,10)
[model_regul_20]=regul(model_chosen, 16,20)
[model_regul_40]=regul(model_chosen, 16,40)

function [model_regul]=regul(model, si, c)

Util=model.U(:,1:si);
Sigtil=model.S(1:si,1:si);
Vtil=model.V(:,1:si);

out      = model.Outputs(:,1:end-1);
inp      = model.Inputs(:,1:end-1);

kapa=[Sigtil*Vtil';inp]*[Sigtil*Vtil';inp]';
W=eye(size(model.sys_red{si}.A));
W(1,1)=100;
WW=[W, zeros(size(model.sys_red{si}.A,1),size(model.sys_red{si}.B,2));...
zeros(size(model.sys_red{si}.B,2),size(model.sys_red{si}.A,1)), ...
zeros(size(model.sys_red{si}.B,2) )];

nrs=[model.sys_red{si}.A model.sys_red{si}.B;...
model.sys_red{si}.C model.sys_red{si}.D]*kapa*(kapa+c*WW)^(-1);

An=nrs(1:si,1:si);
Bn=nrs(1:si,si+1:end);
Cn=nrs(si+1:end,1:si);
Dn=nrs(si+1:end,si+1:end);

model_regul=ss(An,Bn,Cn,Dn,2);

end
```

The location of the poles changes for each of the regularizations. Note that this is achieved by trial and error. The change in the poles' locations can be seen in Figs. 7.9

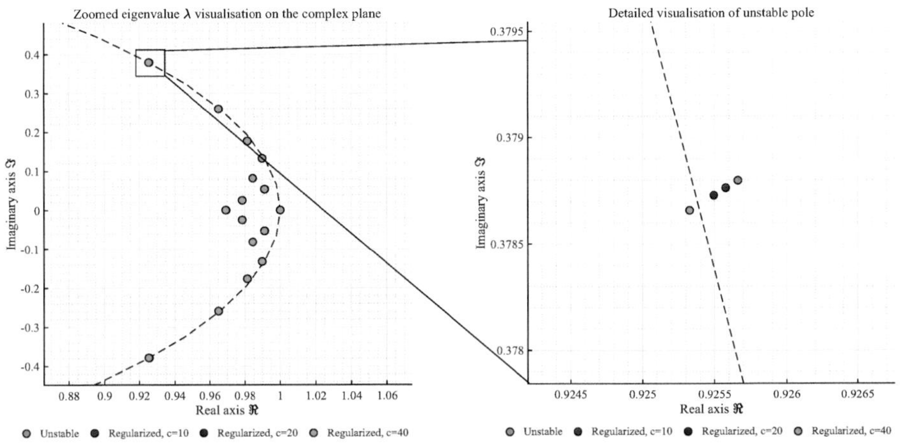

Fig. 7.9 Unstable model pole's location change with relation to unit circle. On the right side, the change of the unstable pole is visible in detail, where the unstable pole is moved inside the unitary circle. The change is barely noticeable given the pre-existing proximity of the unstable pole to the unitary circle

and 7.10. The change is only minimal; however it is sufficient to bring the unstable pole back inside the unitary circle, thus making the model stable. In Fig. 7.9, on the left side, the change in the poles' locations is practically invisible. A zoomed in perspective on the unstable pole on the right side shows that by making use of the regularization technique with the positive scalar value c equal to 40, the pole returns inside the unit circle. In Fig. 7.10 the change in each of the 16 poles modulus of the ROM with relation to the unstable poles modulus is shown. Although the difference is not substantial (differences in the modulus are in the order of 10^{-4}), it can be observed that the change is not the same across all poles. For example, the pole at the origin, the seventh pole, maintains its location.

This chapter focused on two aspects related to the algorithms tested in this book. Given that they concern some nuances related to the algorithms tested, the chapter is labelled Advanced Topics. Two topics were discussed. First, it was shown how the estimation of initial conditions for the linear simulation can affect the assessment of the model and the reconstruction of the wake. It was seen that the RON state vector at the initial time instant can be estimated by projecting the known flow field to the low dimensional space. This option was shown to be preferable to finding the initial state vector by maximizing the fitness of the output prediction and data, which could lead to an improper reconstruction of the wake. The second topic was related to post-processing the ROM to ensure stability, a processed termed regularization. It was seen how the state space model can be tweaked to fully ensure that all poles lie within the unit circle. This process was shown to affect poles differently.

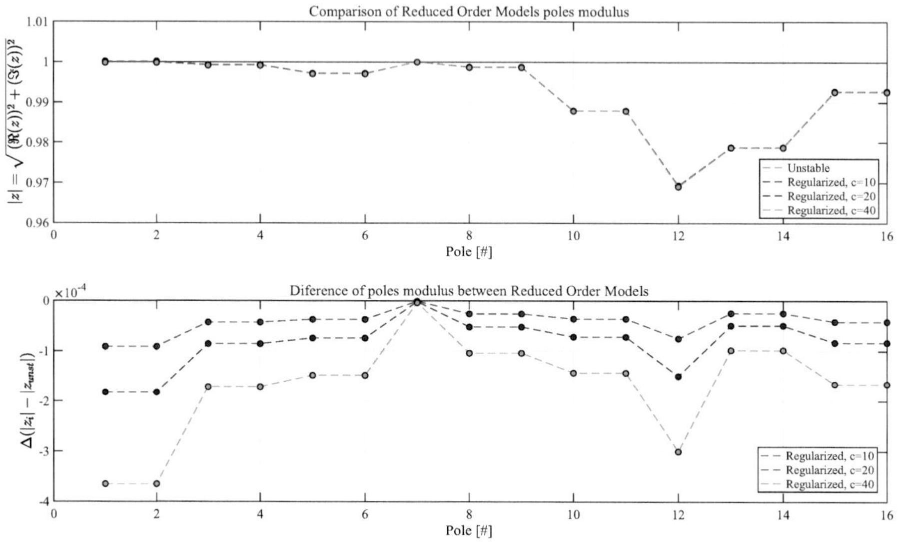

Fig. 7.10 Change in the Reduced Order Model poles modulus depending on the value of the positive scalar c used in the regularization

References

1. T. Van Gestel et al., Identification of stable models in subspace identification by using regularization. IEEE Trans. Autom. Control **46**, 1416–1420 (2001)
2. N. Cassamo, J.-W. van Wingerden, Model predictive control for wake redirection in wind farms: a Koopman dynamic mode decomposition approach, in *2021 American Control Conference (ACC)* (2021), pp. 1776–1782. https://doi.org/10.23919/ACC50511.2021.9482631

Chapter 8
Conclusion

This book set out to provide a comprehensive introduction to various topics that are deeply intertwined and together give rise to the core topic of this book: data-driven modelling of wind farm flow control strategies. On the one hand, it focused on the existent technologies that aim at manipulating the wake interaction between turbines to improve the performance of wind farms. For that purpose, the key principles related to wind turbine control were introduced prior to moving to the overarching goal of controlling the farm. Typical modelling approaches based on physical principles were discussed. On the other hand, it also focused on data-driven modelling algorithms that aim at building simplified models from data only, circumventing the necessity of including knowledge of the underlying physics of the system. The technique which was explored in great depth was the Input Output Dynamic Mode Decomposition (IODMD), along with the variants motivated by the Koopman operator. Furthermore, the initial original algorithm from which IODMD departs—Dynamic Mode Decomposition (DMD)—was also explored.

These two fields can be viewed as separate, as DMD related algorithms can be applied to any scientific area of interest, spanning from fluid dynamics, epidemiology, neuroscience or finance. An incentive must naturally be present. In this case it is the existence of systems of high dimensions and governed by complex underlying physics that motivate using these techniques. In this book we have explored the application of data-driven modelling techniques to the field of wind farm flow control, an epitome of the generalist path. In doing this, it has been shown that combing these two fields has great potential, proving that the end user can create Reduced Order Models (ROMs) that can accurately replicate the highly non-linear behaviour of the wind farm system. In addition, the end user may use these models to further investigate the characteristics of processes that take place in the wake. In other words, the resulting models can be interpreted, which adds value in increasing the understanding of the occurring physical phenomena.

The purpose of this book is mainly educational, hoping to facilitate the transfer of knowledge in these exciting and growing fields. The interest in wind farm flow

N. Cassamo, J.-W. van Wingerden, *Data-driven Modelling of Wind Farm Flow Control Strategies*, https://doi.org/10.1007/978-3-031-84116-3_8

control technologies grows as its potential to increase energy production of wind farms increases. The interest in data-driven algorithms is also expanding, as the availability of data and processing power becomes widely available. Furthermore, with the basic principle in mind that the best learning approach is hands-on through experimentation, the book provides code snippets and the datasets so that the reader can replicate the results. The reader is also invited to explore different variations of the techniques shown. Many of the results throughout the book are based on certain preprocessing steps that can be varied (e.g. how the flow field to be used in the data matrices is resampled, the choosing of the observables in the snapshot data matrices, the removal of trends or scaling of input-output signals and the choosing of the input-output data itself, only to name a few).

The reader should bear in mind the caveats and limitations of the results and implementations showcased throughout the book. This initial research and proof of concept is best performed in a simplified scenario, which explains the simplistic simulation set-up. A uniform inflow profile was used. Turbulence intensity and surface roughness were set to zero, and the Atmospheric Boundary Layer stability was considered to be neutral. These settings are obviously not representative of realistic conditions; however they provide an excellent baseline case to explore many of the algorithms discussed. As the complexity of the simulation conditions increases by inclusion of turbulence or other phenomena, it can be expected that the turbine's response becomes increasingly difficult to model. Therefore, it may be the case that the ROM accuracy decreases. However, in an optimistic and hopeful note, the modifications motivated by the Koopman operator offer a possible route to successfully embed the non-linear dynamics in a linear state space model. In addition, it should also be emphasized that the simulations only covered a specific range of operating points and that the models evaluated are only valid within the vicinity of such operating conditions. This is a known limitation of linear data-driven models, although techniques related to Linear Parameter Varying (LPV) models may offer a possibility to have models suitable for a wider range of operating points.

The book has had as main focus the task of modelling. In fact, going back to the main applications of Dynamic Mode Decomposition—diagnostics, state estimation and control—the book focused on the first two. The techniques used aimed at building reduced order models (state estimation) and extracting the key spatio-temporal patterns that help in explaining certain behaviour in the wake propagating downstream (diagnostics). Still, the third application was not explored.

However, the reader should not be discouraged of making use of these models for control purposes. As it was briefly discussed, this can be a challenging application, as it requires the linear model to predict the future behaviour of a non-linear system. Still, for certain operating conditions and specific time intervals, the model may be useful to help in designing a controller for the system. A small example involving the use of a reduced order model to design a model predictive controller for wake steering using a two turbine farm in a simulation environment was briefly discussed and can pave the way for developments which suit conditions of increased complexity.

In addition, current breakthroughs in the existing algorithms may be able to improve the accuracy of the underlying reduced order model. For example, physics-informed Dynamic Mode Decomposition (piDMD) is a recent variant of the DMD algorithm. The latter requires certain physical laws to be respected by the admissible models in the solution of the problem. The algorithm piDMD is mentioned to outperform the classical DMD in certain metrics for specific problems studied in [1].

Throughout the book, as different topics were explored, references were added. The goal is that the reader can deepen her or his knowledge by deep diving into specific literature pertaining to a certain knowledge domain. This can be related to, for example, recent developments in engineering wake models to predict turbines' power production, campaigns to test wind farm flow control technologies in the field in order to increase their technological readiness level or different applications of DMD related algorithms.

I hope you have found the contents of this book useful and at this point have become more knowledgeable in this field. During this process, I also hope you enjoyed learning based on the available datasets, experimented with the algorithms and tried new variations that fit your research interests.

Reference

1. P.J. Baddoo et al., *Physics-informed dynamic mode decomposition (piDMD)* (2021). arXiv: 2112.04307 [math.DS]

Correction to: Data-driven Modelling of Wind Farm Flow Control Strategies

Correction to:
N. Cassamo, J.-W. van Wingerden, *Data-driven Modelling of Wind Farm Flow Control Strategies*,
https://doi.org/10.1007/978-3-031-84116-3

A spelling error in the name of the volume author has been corrected after the initial publication from Jan-Willen to Jan-Willem.

The updated version of this book can be found at
https://doi.org/10.1007/978-3-031-84116-3

N. Cassamo, J.-W. van Wingerden, *Data-driven Modelling of Wind Farm Flow Control Strategies*, https://doi.org/10.1007/978-3-031-84116-3_9

Appendix A

A.1 Linear Algebra

l_2 norm The l_2 norm is a standard method to compute the length of a vector, $\boldsymbol{x} = \begin{bmatrix} x_1 \ x_2 \ \cdots \ x_n \end{bmatrix}^T$, and is defined as the square root of the sum of the squares of the values in each dimension:

$$|\boldsymbol{x}| = \sqrt{\sum_{k=1}^{n} |x_k|^2} \tag{A.1}$$

Frobenius Norm The Frobenius norm, or Hilbert-Schmidt norm, is a matrix norm defined as the square root of the sum of the absolute square of its elements [1]:

$$\|\mathbf{X}\|_F = \sqrt{\sum_{i=1}^{m} \sum_{j=1}^{n} |a_{ij}|^2} \tag{A.2}$$

Singular Value Decomposition
The Singular Value Decomposition (SVD) is a matrix factorization whose computation is a step in many algorithms. The SVD of $\mathbf{A}$ is a factorization:

$$\mathbf{A} = \mathbf{U}\boldsymbol{\Sigma}\mathbf{V}^* \tag{A.3}$$

where

$$\begin{aligned} &\mathbf{U} \in \mathbb{C}^{n \times n} \text{ is unitary} \\ &\mathbf{V} \in \mathbb{C}^{m \times m} \text{ is unitary} \\ &\boldsymbol{\Sigma} \in \mathbb{C}^{n \times m} \text{ is diagonal} \end{aligned} \tag{A.4}$$

N. Cassamo, J.-W. van Wingerden, *Data-driven Modelling of Wind Farm Flow Control Strategies*, https://doi.org/10.1007/978-3-031-84116-3

U and **V** are unitary matrices[1] such that $\mathbf{U}^*\mathbf{U} = \mathbf{U}\mathbf{U}^* = \mathbf{I}$ and $\mathbf{V}^*\mathbf{V} = \mathbf{V}\mathbf{V}^* = \mathbf{I}$. The matrix $\boldsymbol{\Sigma}$ has entries along the diagonal corresponding to the singular values ordered from largest to smallest, which are assumed to be non-negative. The columns in **U** are referred to as the left singular vectors and the columns in **V** as the right singular vectors.

This procedure produces a hierarchical matrix decomposition which spits a matrix into a sum of rank-one matrices given by the outer product of a column vector (let singular vector) with a row vector (conjugate transpose of right singular vector). These rank-one matrices are ordered by the singular value so that the first r rank-one matrices form the *best* rank-r matrix approximation of the original matrix in a least squares sense. For more information on the SVD, as well as the geometric interpretation, the reader is referred to [2].

A.2 Vector Calculus

Gradient

The gradient vector function of a function $f :\in \mathbb{R}^n \rightarrow \mathbb{R}$ at a point $\boldsymbol{x} \in D_f$ is defined as [3]

$$\operatorname{grad} f(\boldsymbol{x}) = \left(\frac{\partial f}{\partial x_1}, \frac{\partial f}{\partial x_2}, \ldots, \frac{\partial f}{\partial x_n} \right) \equiv \nabla f(\boldsymbol{x}) \tag{A.5}$$

Divergence

Divergence is a differential operator applied to a three dimensional vector-valued function $\boldsymbol{F} = F_x \boldsymbol{i} + F_y \boldsymbol{j} + F_z \boldsymbol{k}$.

The divergence of a vector $\boldsymbol{v}$, in Cartesian coordinates x, y, z, is given by

$$\nabla \cdot \boldsymbol{v} = \frac{\partial v_x}{\partial x} + \frac{\partial v_y}{\partial y} + \frac{\partial v_z}{\partial z} \tag{A.6}$$

References

1. G.H. Golub, C.F. Van Loan, *Matrix Computations*, 3rd edn. (The Johns Hopkins University Press, Baltimore, 1996)
2. L. Trefethen, D. Bau, *Numerical Linear Algebra*. English (Society for Industrial and Applied Mathematics, Philadelphia, 1997)
3. S.J. Miklavcic, *An Illustrative Guide to Multivariate and Vector Calculus*. English (Springer, Berlin, 2020)

[1] A complex square matrix is unitary if its conjugate transpose is also its inverse.

Index

N. Cassamo, J.-W. van Wingerden, *Data-driven Modelling of Wind Farm Flow Control Strategies*, https://doi.org/10.1007/978-3-031-84116-3

GPSR Compliance

The European Union's (EU) General Product Safety Regulation (GPSR) is a set of rules that requires consumer products to be safe and our obligations to ensure this.

If you have any concerns about our products, you can contact us on ProductSafety@springernature.com

In case Publisher is established outside the EU, the EU authorized representative is:

Springer Nature Customer Service Center GmbH
Europaplatz 3
69115 Heidelberg, Germany

Batch number: 10756376

Printed by Printforce, the Netherlands